4

Hart, Stephen.

The language of
animals.

$22.50

"Nim Chimpsky" learns the parts of the body by imitating the human teacher's gestures.

OTHER SCIENTIFIC AMERICAN FOCUS BOOKS

Cosmic Collisions

Medication of the Mind

THE
LANGUAGE
OF
ANIMALS

STEPHEN HART

Foreword by Frans B.M. de Waal

A SCIENTIFIC AMERICAN FOCUS BOOK

Henry Holt and Company
New York

Henry Holt and Company, Inc.
Publishers since 1866
115 West 18th Street
New York, New York 10011

Henry Holt ® is a registered trademark
of Henry Holt and Company, Inc.

Published in Canada by Fitzhenry & Whiteside Ltd.,
195 Allstate Parkway, Markham, Ontario L3R 4T8.

Library of Congress Cataloging-in-Publication Data
Hart, Stephen.
The language of animals / by Stephen Hart;
foreword by Frans B.M. de Waal.—1st. ed.
p.   cm.—(A Scientific American focus book)
Includes index.
1. Animal communication.   I. Title.   II. Series.
QL776.H37—1996   95-34585
591.59—dc20   CIP

ISBN 0-8050-3839-6
ISBN 0-8050-3840-X (An Owl Book: pbk.)

Henry Holt books are available for special promotions
and premiums. For details contact: Director, Special Markets.

First Edition—1996

Conceived by Robert Ubell Associates, Inc.
Project Director: Robert N. Ubell
Project Manager: Barbara Sullivan
Art Direction: J.C. Suarès
Design: Amy Gonzalez
Production: Christy Trotter

Printed in the United States of America
All first editions are printed on acid-free paper. ∞

10 9 8 7 6 5 4 3 2
10 9 8 7 6 5 4 3 2   (pbk.)

Vervet monkey.

CONTENTS

Frans B.M. de Waal, Department of Psychology,
Yerkes Regional Primate Research Center,
Emory University, Atlanta

hen we run into an old friend, we part our lips, exposing our teeth. When we are ashamed, the small arteries in our face and neck swell up, advertising our internal state with a bright red color (making it actually worse!) And when mentioning something small we place thumb and index finger together, while we spread our arms when discussing something large.

Most of the time, we do these things without thinking. In fact, we have remarkably little control over them. For example, we are unable to blush on command, and also unable to suppress it when we'd rather not be blushing. And we gesture even when it cannot be seen, such as during a telephone conversation. Our bodily communication is largely automatic.

If language sets us apart from the animal kingdom, nonverbal communication connects us to it. We share a range of facial expressions and gestures with our fellow primates, the monkeys and apes. Young chimpanzees, for example, play with open-mouthed faces that look a lot like laughing. When tickling each other, they utter hoarse, staccato sounds. Although more restrained than human laughter, these sounds are so reminiscent of it that they have the same infectious quality. Thus, I cannot help but chuckle whenever I hear chimpanzees under my office window having fun.

The meaning of nonverbal signals varies with each situation. For example, people smile when happy but also when nervous. We have absolutely no trouble distinguishing between the two; we are masters at handling contextual information. Similarly, a chimpanzee may stretch out an open hand to another in an  attempt to get food. But the same gesture is also used to solicit support: a young chimpanzee under attack by a peer may run to its mom, holding out a hand to her. The mother may then come over to help him out. Or, after a fight, one of the combatants may stretch out a hand to the other. If the other accepts the invitation, the peace will be sealed with a kiss and embrace. So, depending on the context, a single gesture may represent begging for food, a request for support or a conciliatory overture.

Naturally, we expect chimpanzees to have a rather human-like communication with a great deal of flexibility; after all, they are our closest animal relatives. But how

 about the millions of animals, from insects to birds, with whom we have much less in common? Their communication is elaborate, too, and often of an astounding complexity. How these animals attract and repel one another, and synchronize their activities, is no trivial matter. Their survival depends on it. Regardless of the species, the basic premise is always the same—two or more members of the same species need to negotiate an arrangement between them. Be it copulation, taking turns at the nest, or the establishment of a rank-order, they need to communicate their wishes and intentions to each other, and make clear how strong, sexy, or well-intentioned they are.

Without communication, each individual would merely be an island isolated from all other such islands. Communication allows members of a species to come together and coordinate their lives both in a positive sense, such as in the cooperative ventures of honeybees or hunting dogs, and in a negative sense, such as when a male nightingale sings his lungs out to keep rivals out of his territory.

Many signals are species-typical, that is, they develop naturally in all members of a particular species. Thus, all geese threaten intruders by stretching and lowering their necks, and all jackdaws invite close contact with a mate by offering their neckfeathers for preening. Emphasizing their stereotypical and instinctive nature, early ethologists, such as Konrad Lorenz, called these signals fixed action patterns. The implication of innateness posed a serious challenge to scientists who tried to explain all behavior on the basis of learning. It also had serious implications for our view of human nonverbal communication, which is now widely regarded as influenced by both genetic and environmental factors.

In fact, the line between innate and learned communication has become so blurred that most scientists have given up on the distinction, seeing it as a false dichotomy. This is particularly true for communication signals that vary across groups or populations, such as the birdsong dialects discussed in chapter 7. These vocalizations possess universal species characteristics but also vary from place to place—they are at the same time inherited and learned. Clearly, the situation with regard to animal signals is not nearly as simple as used to be assumed.

In chimpanzees, we sometimes even see arbitrary signals of which the origin is unknown, a bit like words in a human language. For example, in the Mahale Mountains, in Tanzania, male chimpanzees invite females with a so-called leaf-clipping display. The male picks up several leaves, then bites them to pieces while pulling the leaf blades away from his mouth. He thus produces a special ripping sound that attracts female attention. This courtship display is known only from the Mahale community. Another gesture typical of the same community is so-called hand-clasp grooming: two chimpanzees clasp each other's hand and raise both hands above their heads, while grooming each other with their free hands.

Hand-clasp grooming is known in only two chimpanzee communities in the wild at far-apart locations. It is common within each of these communities, yet totally absent in nearby ones. This makes it a strong candidate for a "cultural" pattern, that is, a habit transmitted through learning from one generation to the next.

A few years ago, I received a shock of recognition when two grooming chimpanzees here at the Yerkes Field Station suddenly raised their arms with their hands squeezed together. It was exactly as in the few published photographs of hand-clasp grooming. I had never seen anything like it in captive chimpanzees, and knew how rare it was in the wild. One of the two females involved was Georgia. It turned out to be her "invention," as we saw it subsequently only in combinations involving this female. Since Georgia was born and raised in this very group, she could not have brought it from outside.

Initially, we observed hand-clasp grooming less than once a month. Now, several years later, it has become more routine, and is shown by a wider range of individuals, sometimes not involving Georgia at all. On occasion, I have looked out of my window at a "forest" of lifted arms when two or three pairs of chimpanzees were engaged in this strange ritual at the same time. Since it never occurs in any other chimpanzee group that I am familiar with, it clearly sets this particular group apart.

As in our own species, the nonverbal communication of animals thus ranges from the stereotypical and species-specific to the cultural. It is one of zoology's richest topics, and this book provides enthralling examples of the way animals stay in touch with their own kind.

Nim Chimpsky signing "me."

You'd be hard pressed to find a person alive who didn't grow up with talking animals. From 1930s Mickey Mouse cartoons to 1994's *The Lion King*, from *Puss in Boots* to Brian Jacques' mouse hero Mattimeo, children's fiction in every medium abounds with animals talking to their own species, to other species and to humans. And the concept predates both cartoons and books; folk tales from almost every culture contain talking animals.

Children soon learn that animals don't talk. Or at least that humans can't understand them. Or can they? Consider humankind's best friend. First planting a cold, wet nose firmly against its master, it then trots to the door, whining. Message heard and understood: "I want to go out." A cat leaps on the desk at noon and walks back and forth on the computer keyboard, then leads its owner to the empty cat food dish (probably licked clean by the dog). Again, the animal has communicated to the human. (Humans clearly belong to the animal kingdom, but to avoid repetition, I'll often use "animals" to mean "nonhuman animals.")

In their own realm, animals also get their messages across. A male dragonfly swoops down on a female, grasping her for an aerial mating. He rarely grabs the wrong species because body size, shape and color communicate her identity. With a dab of the right color paint, however, a researcher can fool the male. Communicating by way of body shape or color obviously requires no awareness on the part of the sender, no intention to tell a story. Several other types of communication likewise require no consciousness on the part of the sender. Many female animals, from moths to naked mole rats and marmosets, use odor communication so powerful it can draw a male moth from miles away or prevent all but one of the female naked mole rats in a colony from ovulating.

These animals may have no conscious awareness of their body shape or odor. But in many other cases, animals communicate more selectively. A common message appears to be "Hi, I'm a male. Let's mate." Females, attracted by the signal, quietly approach the male. Given only that message, however, females would have little basis on which to choose mates. In most species, the males also appear to add a superlative: "I'm not just a male, I'm a *great* male." Do the females then pick the best advertiser? Charles Darwin addressed this question in his *The Descent of Man and Selection in Relation to Sex*, published in 1871.

Dragonflies.

"There remains a question which has an all-important bearing on sexual selection, namely, does every male of the same species excite and attract the female equally? Or does she exert a choice, and prefer certain males?" Darwin's colleague, Alfred Russel Wallace, who also published on natural selection, insisted that females could not choose mates because no female animal had the mental capacity to distinguish between males. Most scientists studying evolution of behavior today would disagree with Wallace. But how female selection works remains controversial and complicated.

Evolutionary scientists propose several ways a female might make her choice of males—as well as several evolutionary rationales. In some species, the male provides food, help or protection, giving the female an immediate benefit; she therefore chooses the mate she thinks will best provide for her. If she's right, her mate helps keep her healthy, and she produces healthy offspring who will carry her genes and her mate's genes, including, presumably, those coding for being a good provider. As a result, her genes stand a better chance in generations to come. But in other species, the male merely contributes sperm. And here the story gets more complicated.

A female may choose traits that correlate with good genes. She could do this in a number of ways, choosing the oldest male, for example— the one with the largest body, deepest voice, or most finely honed display. These males have survived longer than younger males and so may pass on their survival genes. Or she may choose active or healthy-looking males—the ones who dance

Indian peafowl (peacock).

Springbucks.

most frenetically or sport the most pristine plumage. Healthy males presumably would contribute genes for healthiness. She might choose dominant males—those able to protect the largest, choicest or most highly decorated territory, or those most able to fight off other males.

In all these examples, females trust the signal. But their selection could turn out to be a mistake. Imagine that peahens choose peacocks with longer tails—males who have obviously enjoyed good health in the months it takes to grow a resplendent tail. They might also contribute good genes for a variety of other traits. Female selection for longer tails ends up favoring the evolution of longer and longer tails. At some point, the graph of increasing tail length crosses the graph of increasing male fitness. Beyond that point, tails become a drag on the male's fitness, and no longer advertise genetic superiority. Where long tails once correlated with many good traits, they now become empty advertising. The process has become a runaway feedback system. Most females continue to choose the longest-tailed males, even though a male with a slightly shorter tail represents a better genetic choice. Eventually, of course, the system regains equilibrium, as females that select for other characteristics in making their choice reproduce with slightly better success.

Some scientists speculate that for a signal to indicate desirable genes, it must obviously cost the male something. Called the handicap principle, this idea suggests that extravagant plumage, huge antlers, or fantastic displays actually lower the viability of the male. Only males who can afford such luxury can pull off the display, the theory goes. Cheap displays may not indicate desirable genes, so females look for an expensive mating call, dance or plumage, one that appears of no immediate use to the male. Such costly communications may also signal health to predators. Certain springbucks, small antelopes of the Kalahari desert, communicate their excellent health to hunting dogs or hyenas with a curious behavior called stotting. Instead of running flat out when they see a predator, some springbucks punctuate their escape with soaring leaps in the air. Jumping higher than their neighbors backs, with legs held stiff and parallel like a gymnast's, the springbucks appear to say, "there's no use chasing me, I could obviously outrun you." The handicap principle ensures the honesty of messages, some scientists contend, although a particular individual can still cheat.

The possibility that animals can "deceive" transcends mating communications, spilling over into communication with potential predators, sexual rivals and members of social groups.

Killdeer nest on rocky shores, exposed to predators. The gray and white bird nestles down, neatly camouflaged among the tan grass and gray pebbles. If you happen upon one, you'll see a surprising spectacle. Instead of freezing or flushing, the killdeer trots away at a moderate pace, with one wing held out and down at an awkward, broken-looking angle. Should you pick up the pace and follow—as a hungry raccoon might—you'll find she keeps just out of reach, leading you ever farther from her nest. Turn back toward the nest and she'll circle around and hobble all the more piteously. Finally, when she's led you far enough astray, she rises like Phoenix, suddenly made whole.

Apparent deception does not require even the brain complexity of a bird. Fireflies present a fascinating drama of deception worthy of a mystery novel, complete with plot twists and killers. These small beetles, found worldwide, emit light by mixing chemicals in a special organ in their abdomens. The male of the species blinks out a particular pattern of pulses, watches for the appropriate response from a female of the same species and then homes in on her beacon to mate.

The unwary male lighting next to a sexy-looking flashing light may get a deadly surprise, however. In some species, predatory females mimic the female flashes of other species, luring males to their deaths. These females succeed only about ten percent of the time. Males typically land a short distance from a beckoning female to more carefully assess her flashing code or perhaps to double-check her identity by sensing odors.

Hesitation aids certain males, who appear to add another twist to the plot. They first approach a female of their own species and assure themselves of her identity. Then they begin imitating the flashes of predatory females. This behavior prompts competing males to hesitate longer, giving the first male time to lay a lasting claim to the female target.

These animals may be doing what comes naturally, with no consciousness or intent, but they are not just playing a fixed behavioral tape. The killdeer matches the pace of the supposed predator, and appears to monitor its reactions, overplaying her part if she's not getting the attention she wants. And male fireflies send their own species-specific signal when attempting to communicate to a female and another species' signal when attempting to scare off competing males. These behaviors imply at least a complex unconscious repertoire of communication. While it may be hard to imagine a firefly is conscious of cheating, and it may be stretching a point to consider the killdeer fully aware of faking an injury, some instances of deception among great apes might change the mind of even the most ardent skeptic.

Frans B.M. de Waal studied a large colony of chimpanzees in the outdoor enclosure of the Burgers' Zoo in Arnem, the Netherlands. In his book *Chimpanzee Politics*, he describes several observations of chimpanzees who appeared capable of calculated deception. Two examples:

Hurt in a fight with Nikkie, a chimpanzee named Yeroen began limping badly. But with careful observation, de Waal and his colleagues eventually realized the ape limped only when he was within sight of Nikkie. As soon as he rounded a corner or circled behind the aggressor, his limp mysteriously disappeared.

Another chimp, named Puist, developed the habit of faking a reconciliation gesture. After a fight, a chimpanzee will extend its hand, almost as if offering to shake. When Puist was getting nowhere in a fight, she would sometimes stop, approach slowly and extend her hand. When the opponent reciprocated, Puist would grab her and launch another attack.

Watching chimpanzees and many other animals communicate, an observer cannot help but notice striking similarities between their behavior and ours. Vervet monkeys, for example, give one call when they see a snake and another when they see an eagle. Chimps and other animals communicate disingenuously. Bees, as simple as they are, appear to communicate information about distant nectar sites. Are these behaviors languages? No one equates bee dances or even chimpanzee hoots with human language. Our means of communication, no matter what culture we grow up in, far surpasses in complexity and subtlety that of any animal. But is the difference one of degree or kind?

Historically, philosophers have contended that language sets humankind apart from the simple beasts. Although all modern scientists fully accept Darwin's assertion that most traits blend from simple to complex, scientists fall into two camps when it comes to language. One group insists human language bears little resemblance to animal communication, and resists the use of the word "language" to describe animal communication. These scholars—including many linguists—define "language" using features of human languages such as creativity, rules, and meaning. Another group counters that animal communication can be measured from simple to very complex—such as human languages—just like any other trait. These scientists point to monkey and chimpanzee behavior in the wild, and to experiments in which great apes have learned to communicate to some extent with humans, as evidence that the differences between human and animal languages are differences of degree and smaller than many believe.

No matter how exalted human language, we can communicate with some animals—to a degree. Great apes have learned "languages" based on hand gestures and symbols. Parrots can learn to speak words, and even use those words to demonstrate feats of learning. Certainly a sharp "No!" carries some meaning to a dog. And many animals can learn to respond to hand gestures and voice commands—we call this "training." But another kind of experiment shows humans' ability to convey meaning to animals of almost any kind. By recording animal sounds and playing them back, we can attract animals' attention or elicit the same behavior as the original call. In some cases, researchers have modified signals and elicited modified behavior. These so-called playback experiments represent one of the most powerful tools scientists employ to begin to understand how animals communicate.

Caveats and Confessions

The field of animal communication spans an incredibly wide range, from the color patterns of cuttlefish to the complex social life of dolphins. Every species that reproduces sexually must communicate at least enough so that it can mate. And the kinds of communication employed include all five senses humans pay attention to, as well as some we are unable to detect, such as the electric sense of sharks and the infrasound awareness of whales. I have dipped into this sea of fascinating information and netted a few samples. I've necessarily left more out than I could include, hoping only to whet the reader's appetite to learn more—by reading, to be sure, but even more importantly by observing animals first hand.

Finally, I confess I've made my selections and written the text as an enthusiast. I've groomed a pet South American spider monkey, whistled to pet guinea pigs, monitored pet electric fish on my stereo, said, "Do you want to go for a walk?" to numerous dogs, and discussed the state of the food bowl with several cats.

I grew up loving animals, fictional and real. But I also trained in neurobiology, a science much easier to control than language research and field observation. I hope both my love for animals and my skepticism show through in this book, as well as my firm belief that we can make the world a better place by learning about and conserving the creatures with whom we share this beautiful planet.

Hans, Too Clever by Half

In the late 1800s, a retired German schoolteacher named Wilhelm von Osten set in motion one of history's most infamous demonstrations of animal communication. The experiment, involving his horse Hans, still casts a shadow over studies attempting to plumb the depths of human-animal communication.

Von Osten thought Hans a clever horse, and set out to teach him to perform basic arithmetic. The former teacher determined to use the same teaching methods he had used with his students. He began simply, but soon advanced the problems, surprised Hans could do so well. Hans solved double-digit subtraction problems, stamping one front hoof to count out the answers. He rarely erred.

Hans' fame spread throughout Europe, drawing curious onlookers, including a number of eminent scientists. Few left doubting Hans' skill. He could even answer questions posed by visitors when his trainer was out of sight.

Two doubters, however, determined to devise a test more clever than Hans. First they decided on a problem. One whispered it to Hans, then moved out of his sight. Amazingly, Hans still solved the problem. But when the whisperer gave Hans a problem no one else knew and then left the room, Hans stumbled. His chance of success fell to the level of chance.

Hans was in fact clever beyond anyone's imaginings—but not at math. Instead, he was clever at reading human body language. Von Osten learned to his surprise that Hans simply started stamping and continued until von Osten's body language indicated he'd reached the right number. And Hans had learned to generalize this skill beyond von Osten to other questioners, and even bystanders who knew the answer. But when no one within his sight knew the answer, Hans had no clue when to stop tapping.

The Clever Hans incident chilled enthusiasm for human-animal-communication research for decades. No one today assumes horses can understand human language. But many assert great apes can. In the 1960s and 1970s, attempts to communicate with animals—chimpanzees, orangutans, and gorillas this time—experienced a resurgence. Then the Clever Hans phenomenon struck again. One researcher analyzed videotapes of "conversations" between trainers and his subject, a chimpanzee named Nim, who had learned some word signs derived from American Sign Language. A frame-by-frame analysis revealed trainers unconsciously prompting and modeling each word, with Nim imitating. Unaware of their prompts, the trainers had credited Nim with producing sentences. For the second time, enthusiasm for human-language training for animals waned. The few researchers currently pursuing these studies keep Clever Hans in mind, doing their best to design scientifically "clean" experiments.

Cephalopods

Me Tarzan, you Jane." If the male cuttlefish could speak, this might be his opening line. Although cuttlefish rarely use sound to communicate, the male still has an opening line, and it might translate to the longer, but conceptually simpler, "Me Tarzan. You Tarzan? No? Must be Jane." Cuttlefish and squid communicate using their remarkable ability to control the pigment in their skin. They flash messages in colorful spots, splotches and background color. Cuttlefish add to their unique visual communication certain swimming postures and gestures of their ten tentacles. Along with octopuses, cuttlefish and squid belong to the class Cephalopoda, molluscs like snails, slugs and clams. Cephalopods, mental giants of the mollusc world, manipulate objects with tentacles, swim with jet propulsion, eat with beaks and see with eyes as complex as ours.

Direct connections from the brains of cephalopods to special muscles allow split-second change in skin color by relaxing or contracting chromatophores. These skin-surface cells, filled with red, yellow and black pigments, can change from being spread out to being tightly contracted in a few thousandths of a second. Under the surface layer, white pigment cells and even deeper green cells reflect light when unmasked by contracted chromatophores. Cephalopods can also change their skin texture to enhance communication, raising or smoothing warty-looking bumps. Even though cephalopods appear unable to see colors, they seem to match their surroundings remarkably well.

When not fading into the background, some squid and cuttlefish can create dramatic patterns covering either their whole bodies or only parts of it. In some species, observers have catalogued 31 full-body patterns and calculated a potential repertoire of nearly 300 combinations of full-body patterns, partial-body patterns, skin texture and body posture.

Common European cuttlefish.

The mating dance of the opalescent squid.

Octopuses remain solitary except when mating, and researchers have so far seen little they would call complex communication among them. But like squids and cuttlefish, octopuses do exhibit color changes based on internal physiological states. Males of some octopus species sport enlarged suckers, used in a "sucker display," presumably designed to communicate their sex. Females of one species develop luminescent cells, circling their beaks like green lipstick, that may attract males.

Phosphorescent squid.

Jane cuttlefish—like females of any species—won't be satisfied with just any male. She wants a healthy, vigorous Tarzan whose sperm will carry genes that enhance her offspring's chance to survive, mature, and breed again. So she looks for a number of attributes. Size signifies health, of course, but in addition, cuttlefish and squid who swim with their arms erect and their skins flashing apparently look healthy to females.

Cuttlefish and squid make great food—not just as sushi, but to several oceanic predators—so they normally blend into the background with a mottled, cryptic color scheme. But for the male cuttlefish, when it comes to mating, the chance of passing on his genes outweighs the risk of becoming a meal.

Stretching his arms forward, bunched together or arched into a ten-stranded basket, he flashes a striking zebra pattern, signaling his sex. Other cuttlefish nearby get the message. Males return the salute, but females remain mottled. The absence of the male pattern, rather than any distinguishing features of the female's pattern, tells the male her sex. If a male fails to respond with a zebra pattern—perhaps because of illness—other males may mistake him for a female.

All males in a group strut their stuff with a zebra pattern, and most females remain mottled. But if a nearby female changes from her cryptic mottled pattern to a more uniform gray, she's signaling her readiness to mate. Now the competition between males grows intense, in some species escalating into physical contact and biting. Finally, all males but one—usually the largest—literally turn tail and retreat, shifting back to their normal, unisex mottled pattern—behavior that resembles the submissive posture of a dog with its tail between its legs.

After discouraging nearby males with his prowess, the victorious male turns from aggressive to sensitive. He approaches the female and turns from visual communication to tactile, gently stroking her between her eyes and arms. At first, she may indicate her alarm by flashing an acute disruptive pattern. The male calms her by blowing water at her and jetting gently away. He approaches again and again until the female accepts him, literally with open arms. If a boorish rival should attempt to intrude, the mating male again flashes an intense zebra pattern. If he's swimming side by side with the female, he can even display his stripes only on the side of his body facing the intruder. At the same time, he can maintain his sexually suggestive uniform gray on the side facing the female. At last, the pair links arms and begins to mate. Both now adopt the cryptic mottled pattern that attracts least attention.

Squid, which are more social than cuttlefish, also communicate courtship with skin color. They gather in groups of 10 to 30 individuals, but soon break up into courtship parties of one female and two to five males. The largest male attempts to guide the female away from other suitors. The couple engages in precopulatory mutual rocking, jetting gently to and fro together. If the male approaches too closely at this point, the female may streak away. The male follows, and this teasing game can continue for up to an hour at high speed, possibly representing an attempt on the female's part to assess the male's health. Male squid use a zebra stripe not unlike the

Complete mass of squid egg capsules.

cuttlefish's to ward off other males. They also exhibit a one-sided smooth silver pattern signifying "keep away." The male only displays this lateral silver pattern to other males, keeping the side facing the female sexually stimulating.

Squid don't embrace to mate. Instead, the male merely tries to attach a small, sticky packet of sperm to the female's body. As he reaches out with the sperm packet, he displays a pulsating pattern of chromatophores. If the packet sticks, the female places it in her seminal receptacle, completing the mating ritual.

The social cephalopods, squids (such as *Sepioteuthis sepioidea*) and cuttlefish, clearly communicate internal states—readiness to mate, sexual identification and the like. Human equivalents might be blushing, stuttering and shy body postures. Do cephalopods communicate more than sexuality? Some scientists suggest that their full-body patterns also act as nouns and verbs and small spots and patterns as adjectives and adverbs. Posture and movement might add context. "It could be that if *Sepioteuthis* puts stripe on the side of the body, and then puts golden eyebrows over the top of the eyes, and raises the arms, that it has modified stripe by the golden eyebrows and by the arm raise to mean something more complicated or maybe even different from whatever stripe means by itself," says Jennifer Mather. Mather, a psychologist, studies cephalopod behavior and teaches at the University of Lethbridge, Lethbridge, Alberta. Mather's hypothesis, although intriguing, remains unexplored.

To investigate cephalopod visual communication further, Mather and others would like to "speak" their language. By mimicking the visual cues with a colored model—communicating to the cuttlefish in a sense—researchers could watch for behavior changes and begin to understand their complex communication.

"I would suspect that cephalopods are not going to have a language anything like as complicated as ours by the time we know whether they have a visual language," Mather concludes. "But I suspect we are going to find an interesting communication system when we finally have the time and the energy and the resources to find out."

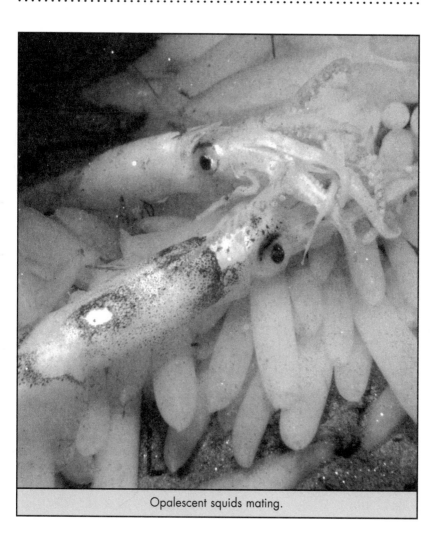

Opalescent squids mating.

Honeybee Dances

oneybees provide humans with more than natural sweetener and sayings such as "busy as a bee." They also provide a rare example of an animal communicating information about objects that are not close at hand. Some say that honeybees, as simple as they are, provide the only example of such symbolic communication among nonhuman animals.

While most insects limit their social lives to quick mating, bees and some of their relatives, as well as termites, form societies with division of labor and cooperation. Cooperation in most species requires some kind of communication, and the honeybee dance provides the archetypic example. A single worker bee (all workers are female, by the way) finding a rich source of nectar flies back to the hive. A short time later, dozens of fellow workers make beelines to the nectar site.

How do the recruits find the site? That question has plagued naturalists since Aristotle. In the early 1900s, Austrian biologist Karl von Frisch began to study honeybees in earnest. At first, it appeared to von Frisch that bees merely sought out the odor inadvertently brought back to the hive by scout bees. But he also noted intense activity among returning bees and began watching closely. Those returning from his sugar-water feeding stations near the hive marched in a busy circle on the comb. Those carrying pollen from distant flowers danced a figure eight.

Von Frisch first called these behaviors the nectar dance and the pollen dance. But on further study, he formed a new idea—busy activity of scout bees after returning to the hive communicated more than just excitement or information about food type. By moving his feeding station farther and farther from the hive, he determined that when the station reached 50 to 100 yards from the hive, nectar gatherers began dancing the figure eight pollen dance. This dance, von Frisch concluded, related to distant food sources, not food type.

The bees bustled around in a purposeful manner, buzzing their wings and waggling their abdomens. They would hootchy kootchy in a straight line, then circle back to the beginning—first circling left, then right. Von Frisch

Dancer bee and followers.

observed that the speed of the waggle and the angle of the line communicated both distance and direction to workers crowding around the scout.

On a horizontal hive, a bee can merely crosscut her circle in the direction of the flowers she found so full of nectar. The speed of her waggle and the number of circuits per minute indicate distance from the hive. But most bee hives consist of vertical combs. How does a scout point out the right direction? Instead of indicating direction from the hive, she indicates an angle from the location of the sun. Forty five degrees to the right of vertical means forty five degrees to the right of the sun. Scout bees even manage to account for the apparent movement of the sun throughout the day. One thing bees cannot account for, however, is reorientation of their hive. Bees normally nest in trees, so it makes no sense for them to evolve the ability to account for a rotated hive. But bee researchers can easily rotate a hive in a box so that it opens in a different direction. In that case, the bees become confused and cannot follow the direction information contained in the dance.

Von Frisch's suggestion, that bees have a "language," aroused the skepticism of other scientists, some of whom continue to this day to favor von Frisch's earlier idea that scent alone guides bees. Nonetheless, von Frisch shared the 1973 Nobel Prize, indicating wide acceptance for his bee-communication research.

A crucial test for any communication system is to modify it experimentally. Could we communicate with bees, using their dance to indicate distance and direction of a target the bees had never visited? In 1989, a team of European scientists led by Axel Michelsen of Denmark's Odense University did just that. They built a bee of brass and beeswax with a bit of razor blade to represent the wings. Their robot bee could buzz its wings at the requisite 280 cycles per second, waggle its bottom and dance in a circle. It could even deliver drops of sugar water to the dance watchers.

Using a computer to control the robot, researchers set their robobee to dancing, attempting to illuminate the components of the dance. Dancing a normal waggle dance, the mechanical bee, like any robot, was a bit clumsy. But it got the message across. Some recruits used the information to fly to a target Michelsen had set up in a field. On a different day, Michelsen sent the bees in the opposite direction. Bees found new targets, regardless of wind direction, correctly following the robot's message. Robobee could not have communicated with odor, because it never left the hive. Despite its clumsiness, Michelsen says, "In our work with the robot bee, the dancer never visited any of the places advertised, and yet the bees turn up at the places indicated." After years of research with robot bee experiments and other studies with individually numbered bees, all but a few scientists are now convinced that the dance does indeed communicate distance and direction, just as von Frisch suggested.

But how do bees perceive the dance? Since hives are dark, the workers crowding around a dancer cannot see. Michelsen says research has shown that it is unlikely that bees can feel the hectic dance through their feet. Michelsen suspects bees sense air currents set up by the scout's wagging backside and buzzing wings. The robot bee also moves its wings and abdomen fast enough to mimic natural air currents. By modifying the dance, Michelsen highlighted the wagging portion of the dance as the key to conveying direction. Stereotyped semicircles, returning the bee to the starting point, may help to orient surrounding bees to where the next waggle will occur. Communicating distance proved more difficult for robobee. Michelsen thinks his robot lacked subtlety in conveying cues for distance, perhaps even giving conflicting information. Michelsen's present research consists of investigating how air currents generated by dancing bees convey information.

Other researchers have shown that only a small number of bees, who keep their heads very close to the dancing bee, get the message. Workers farther away must wait their turn to move in close enough to read the dance. Further research has strengthened Michelsen's supposition that bees using their antennae "hear" the air vibrations around the dancing bee. Without both antennae intact and functioning, bees could not interpret the distance and direction information conveyed by the remarkable dance.

A honeybee entering its hive.

Other Insects

Mosquitoes

A porch swing on a summer evening: the hinges creak, crickets trill from the field, the wind sighs gently. Gradually, you become conscious of a vaguely annoying sensation.

Insects among the heather.

At first, the high-pitched whine barely registers, but it grows louder and closer. Suddenly, the whine appears right in your ear. To you, the sound communicates irritation, like fingernails on a chalkboard. To the male mosquito, it's a beautiful song—the sound of a female mosquito looking for an evening meal of blood.

Male mosquitoes hear and obey the siren call of the female, flying directly toward the sound. "In fact, you can even attract males with a tuning fork of the proper frequency," notes Marc J. Klowden, a mosquito behaviorist at the University of Idaho. Because the sound comes from the female's beating wings—mosquitoes cannot vary their wingbeat frequency at will—it rises in pitch as the air temperature rises. Male mosquitoes detect sound not with ears, but with their antennae, which resonate only at the particular frequency emitted by the female. "The male mosquito antenna is built like a tiny tree sitting on a very small joystick; the branches pick up the vibrations and cause the underlying joystick to move. Movement of the 'joystick,' called Johnston's organ, is translated into nervous impulses by sensory receptors," Klowden explains. The impulses signal the male mosquito's brain, which interprets the sensation as sound.

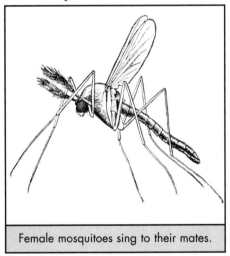

Female mosquitoes sing to their mates.

Fortunately for mosquitoes, the resonant frequency of the male's antennae also rises with temperature, remaining locked in on the pitch of a female of his species. With the mechanical relationship between the female's signal and the male's response, this sexual communication serves its purpose admirably, producing plenty of mosquitoes—many more than we like.

Crickets

The symphony of cricket trills gracing grassy fields in many parts of the world arises from an all-male orchestra. Each male advertises his presence and prowess by scratching together the bases of his forewings, which are ridged like tiny washboards. Each stroke creates a single chirp, and the cricket's song consists of a series of chirps called a trill. Like the communications of many males, the trill carries two meanings, a come-hither message for females and a go-away message for competing males.

To human ears, the field may sound full of crickets all calling at once, but males take turns. If two males call at the same time, one will fall silent and move a respectful distance away. A female, on the other hand, homes in on the most robust trill, with mating on her mind. She will not respond to just any male cricket. The male must sing a song specific to her species. This discrimination is so specific, one scientist found, that female hybrids prefer the song of identically crossbred males. When scientists crossed two species—let's call them A and B—females of species A responded to A males only. B females responded to B males

Brown cricket singing.

only. Hybrid females respond to hybrid calls. By recording the songs and making graphical representations, the scientists found subtle differences between the songs of males with an A father and a B mother (AxB) and males of the opposite cross (BxA). Hybrid females detected the difference as well. Females of AxB crosses preferred AxB males over all three other possibilities, (A, B, and BxA). Scientists explain this amazing pickiness by suggesting that the nerve cells of cricket brains contain a pattern generator. In males, the genetically determined generator makes them sing their species-specific song and no other. In females, the same pattern generator gives them an internal ideal with which to compare any song they hear. Although genetically programmed to play a particular song, male crickets can determine when, where and how loud to play, unlike female mosquitoes, where the signal results directly from beating wings.

A successful male cricket attracts a female of his species, mates and passes his genes on to the next generation. But in an insect version of *Fatal Attraction*, he may inadvertently attract a completely different kind of female. This female is looking not for sex, but for a place to lay her eggs. She's a parasitic fly of the genus *Ormia*. Once she's found a male cricket, *Ormia* lays an egg. The developing maggot burrows into the male and grows, eventually eating him from the inside out. But how can *Ormia* home in on the song of a species so different from her own? *Ormia* is a true fly, the size of a housefly and related to mosquitoes. And like mosquitoes, *Ormia* flies detect sound as their antennae sway, buffeted by the movement of air molecules. In a sense, a female *Ormia* has the same goal as the female cricket—to hear and home in on a male cricket. To succeed, she must hear sounds from a much greater distance than the maximum distance at which her antennae are effective.

Over millennia, flies in the genus *Ormia* have evolved the same solution as crickets—an eardrum. Crickets hear with a small eardrum on their legs. Unlike other flies, *Ormia* females possess the same sort of structure. *Ormia's* evolution has produced a more sensitive detector than the cricket's. Scientists measuring the output of the nerve cells connected to the eardrum of one *Ormia* species found the flies a hundred times as sensitive to male cricket songs as were female crickets.

In places where the parasitic flies and crickets coexist, such as Hawaii, cricket evolution must balance the desirability of mating with the fatal possibility of becoming a meal for a growing fly maggot. The presence of the parasite has caused Hawaiian crickets to develop a shorter chirp. The males also sing only when the flies are least active, during the hours of darkness. In this way, they attract female crickets, but mostly avoid attracting female *Ormia* flies. Crickets of the same species on an island without parasitic flies have a fuller call and begin calling at dusk, persisting until dawn.

Katydids and Bats

Crickets are not the only insects that inadvertently signal to another species. A close cousin of the field cricket, the greener, longer-legged katydid, faces a similar challenge. Instead of attracting the female he bargained for when he launched into full voice, a male katydid may find himself becoming a meal for a bat. A repetitive staccato trill made up of many frequencies is easy to locate, no matter what species produces it. This is, of course, the purpose of the male katydid's trill. But in bat-infested Panama, katydids change their tune. They sing a higher-pitched song, with a narrower frequency range, much harder to locate. They also sing a lot less. In an experimental situation, loud, enthusiastic katydids caged with hungry bats survived less than a minute. Shyer, quieter males lasted more than half an hour before becoming bat bait.

But if males must remain quiet to survive, how can they attract Ms.

The ant queen is approached by a much smaller worker.

Right? A quiet, infrequent trill can bring a female into the vicinity, perhaps to the same plant. The quieter males then complete their attractive act with a silent dance so enthusiastic they shake the leaf they're on. Females detect the dance through the plant, locating the source of both the beautiful voice and the "swiveling hips."

Caterpillars and Ants

Communication between species need not threaten the sender. Certain butterfly caterpillars signal their presence with vibration, as katydids do. They rub special organs, called vibratory papillae, across a rough part of their exoskeleton. This creates an acoustic signal, similar to rubbing thimble-clad fingers across a washboard in a jug-band. But the caterpillar aims its vibratory signal not at another member of its own species—caterpillars are immature butterflies and don't mate—but at ants. These signals call the ants, which then gather sugary secretions the caterpillars produce from special glands on their backs. Caterpillars also produce an odor that alerts ants to predators. In return for the food, the ants act as bodyguards, swarming around the caterpillar and keeping it safe from wasps and other predators. In some species, ants carry the caterpillar into their nest, feeding it there. The chances that a caterpillar would survive without the help of ants are near zero.

Caterpillars may use vibration to call ants to come close to them because ants use vibration to call ants. Many ant species use vibration to communicate within the colony and to call for help when necessary. In one species studied by scientists, a victim of a collapsed tunnel buzzed the ground to attract rescuers from its colony, who dug it out. Other species use vibration to notify their mates of location of food sources. Caterpillar vibrations may attract ants' attention, leading to the discovery of the nutritious secretions.

The ability to signal a different species does not represent a mere quirk of evolution. Two other groups of butterfly caterpillars that associate with ants also call their companions by vibrating the substrate. Neither uses vibratory papillae, however, and scientists do not know how they produce the signals. In contrast, several groups of caterpillars that neither produce sugary secretions nor associate with ants do not send these buzzing signals.

Stoneflies

If katydids were to represent the insect world in disco contests, stoneflies would play the drums in insect rock bands. Members of the grasshopper and beetle orders drum, and ants and other insects communicate by vibrating the substrate, but no arthropod comes close to the

Stonefly.

rhythmic virtuosity of stoneflies. Unlike crickets and katydids, male stoneflies actively search out their mates. A male of one species makes the first move, a brief crescendo of dull taps: "ba da da Da DA DA dum." The female responds with a comparatively quiet "ba da da da da dum." Another species marches to a different drumbeat: "ta TA TA TA TA," and the female responds with a slightly spaced-out "ta-ta-ta-ta-ta." Some species repeat this two-part pattern only. Others use a three-part pattern—male calls, female responds, male confirms—and some use even more complex patterns of call and response. Stonefly specialists have studied about 150 species, each with its own pattern. Lacking drumsticks, stoneflies use their abdomens, tapping or rubbing the ground, or merely shaking their bodies to vibrate the substrate. Some species sport special abdominal appendages for drumming.

The number of beats, the shape of the beat—that is, how sharp the sound is—the interval between drumbeats and the evenness of the rhythm all contribute to the specificity of the song. Some species favor a two-beat-per-second rhythm, others up to 20 beats per second. The rhythm appears to arise, as in crickets, from genetically determined nervous-system wiring.

A computer-generated "drum solo" gets the attention of a female, but only if it comes close to matching the species-specific rhythm. Vary the program too much and the female ignores it. Stoneflies of a single species living in Alaska and Colorado beat out and recognize a slightly different rhythm. The existence of dialects in these two populations suggests to scientists that they may be in the process of evolving into separate species.

In laboratory tests, males and females can converse over almost nine yards of wooden rod and through different twigs of the same limb. When researchers give the two stoneflies separate drumheads— for example, a paper cage in the laboratory, a dry leaf or dry bark in the wild—males and females could only communicate over about two yards. On a solid rock surface, with no drumhead, they couldn't communicate at all.

Only virgin females respond to male drumming. Once the conversation begins, the couple keep at it until the male finds his bride. The more vigorous their duet, the faster the male finds the female and the sooner they can mate. A typical male search in a horizontal laboratory arena might go like this: The male wanders randomly, calling occasionally until he hears a "come-hither" response and begins the duet. He walks a short distance to his left and calls again. He notes the direction of the response and turns sharply right, calling again after aiming in the direction he thinks he should travel. Within a few turns, he finally finds the female. No one has studied male stonefly searching in the wild, where a typical "arena" would consist of many twigs of a highly branched tree. Scientists think the male may make some sort of innate triangulation to locate the female, and that she could judge his fitness as a mate by how long it takes him to find her. To facilitate the search, she stands her ground while communicating, but may move after a while to avoid slow males—and spiders eavesdropping on her conversation. If the male's search proves successful, the pair mates immediately. Better communicators presumably stand a better chance of mating and laying eggs before a predator finds them. And passing genes on to the next generation is, after all, the name of the game, no matter what kind of communication is used.

CHAPTER FOUR

Spiders

A male spider of a web-spinning species faces a dilemma: How can he convince a female on her web that he's a mate, not a meal? The males of many spider species succeed by plucking a mating song on the female's web strings. First, the male must announce his presence—knocking on the door, as it were. If he gets no response, he'll keep trying for days, sometimes waiting until the female completes her last molt and becomes sexually mature.

When he finally receives his invitation—a series of twitches and jerks of the web from the female—he approaches carefully. Because web-spinning spiders see poorly, the male continues his plucking patter right up to mating. Each species has a specific way of announcing and approaching, so the female knows the vibration comes from a male of her species.

Or so she thinks. Certain spiders, called pirates, creep onto the web of a waiting female of a different species. There they vibrate their bodies and twitch their legs in a precise imitation of a male seeking to mate. When the fooled female approaches, the intruder turns the tables and eats her. Here the communication has gone awry for the receiver; the sender gains a meal. Mimicking spiders, which are found in a variety of species, can also mimic the motions of a struggling insect caught in the prey spider's web. Despite the fact that scientists call this "deceptive signaling," pirate spiders have no conscious intent to deceive, and although they can tailor their tactics depending on the species under attack, they can only prey upon a small range of species.

Other species, such as the Mexican *Cupiennius salei*, spin no webs, but still sense the world through vibrations. Like its web-spinning relatives, *Cupiennius* sits and waits for prey to pass by. It emerges from its banana-leaf hideout only at night. How then can it "see" its prey? The final segment of the spider's leg acts a little like a shock absorber, bending slightly to take up vibrations of whatever substrate the spider rests on, then signaling the brain. An insect walking by inadvertently signals its presence with a signature vibration. To the spider, this says "dinner time."

The small male black widow beats a hasty retreat from the female.

Wolf spider.

Detecting the dinner bell among a cacophony of vibrations—rain drops, large animals, even cars and humans—is no easy task. Scientists think the spider depends on the precise frequency of vibrations. Cars shake the leaf at a low frequency, cockroaches at a much higher one. The spiders detects food in somewhat the same way we distinguish a dinner bell from the rumble of a passing bus.

With such a finely honed sense of vibrational frequency, it makes sense that the spiders can detect their mates. But *Cupiennius* carries vibrational signaling far beyond mere mate detection. If a male catches a whiff of female scent, he begins a frenetic courtship dance, shaking his backside like an eight-legged Elvis. He sends vibrations, a sort of love poem, throughout the leaf in pulses punctuated by pauses. A female politely "listens" to the entire poem before responding with a short shake, turning on the porch light, as it were. If the poem is foreign—the product of a different species—the female remains silent. By recording the vibratory love poems of half a dozen closely related *Cupiennius* species, scientists detected differences in the lengths of pulses and pauses. Playing back the vibrations to a female, scientists never saw her respond except to the poem of her own species.

Jumping spiders and wolf spiders don't weave webs, relying instead on their excellent vision to capture prey. Males carry on a courting dialog using vibration and, in some species, by waving their arms in a sort of sexual semaphore. Scientists studying the jumping spider wondered what would happen if a female watched a video of a male courting. Many animals—pigeons and bees, for example—see our 60-frames-per-second television as a confusing slide show rather than a smooth movie. But wolf spiders, it turns out, not only perceive video on even tiny TV screens smoothly, they act like it's real. University of Cincinnati behavioral ecologist George W. Uetz and his colleague David L. Clark showed videos of potential mates to jumping spiders. The spiders returned the sexual signaling, waving their legs and rotating their bodies.

Recently, Uetz studied two closely related wolf spider species. One, called *Schizocosa rovneri*, uses only vibration to communicate its courtship messages. Females of this species ignore leg waving. The other, called *Schizocosa ocreata*, uses both vibration and semaphore. Although the two species live in the same forests, their microhabitats differ. Semaphore spiders live in loose leaf litter, which dampens their vibratory signals. Loose leaves don't affect light, however, and the leg waving apparently allows the male *ocreata* to send his message over longer distances. To enhance its semaphore ability, the species sports small flags—tufts of hairs—on the forelegs to make them more visible.

"We wondered if the visual courtship displays and tufts on the forelegs of *Schizocosa ocreata* evolved as a means of increasing the detectability of their signal." Uetz says. So he and a student shaved the males' legs to see if the female would still respond. Shaving slashed the male's success rate in winning mates only when the female could see, but not hear, him. Shaving alone, however, may have changed the behavior of the males, Uetz worried.

Sending spider signals using video gave Uetz and his colleagues a new approach. Using techniques borrowed from high-tech Hollywood movies, Uetz digitally shaved a virtual spider's legs, then presented the video to a female of the same species. She failed to respond, clinching the conclusion that females pay close attention to the semaphore.

Then Uetz tried one more manipulation. He made videos of male *Schizocosa rovneri* spiders—the ones that don't semaphore while courting—waving their forelegs and showed the tapes to females of the same species. As expected, the females showed no interest. Then Uetz digitally added tufts to

Georg Uetz (left) and colleague Will McClentock with digital spider.

virtual *Schizocosa rovneri* forelegs and played the videos to females. This time they responded. Uetz guesses that the semaphoring *ocreata* split evolutionarily from the older *rovneri*, evolving semaphore signaling to compensate for the move to the loose-leaf-litter habitat. "Using the video method, we are able to test this hypothesis about how evolution might have happened by adding traits to a species that doesn't have them....The visual courtship displays exploit the female's sensory bias (visual prey-detection mechanism), allowing males to identify themselves; the tufts make their message easier to detect," Uetz concludes.

The Case of the Missing Perfume

For a female Sierra Dome spider, a web serves as a safe haven from predators and a way to catch a meal—until her last molt. When she sheds her skin for the last time, the female becomes sexually mature and her web becomes a trysting bed.

If the female dome spider lives in a dense population—a spider city, so to speak—she'll have no need to attract a mate actively. Many males visit her web, testing her mood by touching her body. Most of the time, she shoves them away. Just before she molts, however, the female signals her imminent maturity by ceasing her rebuffs. One masterful male then begins a vigil, spending days fending off rivals. The male who holds the fort when the female finally molts will mate with her and fertilize most of her eggs. All this jousting gives the male a mighty appetite, and he eats most of the prey caught in the female's web. Fortunately for her lover, the female fasts before molting.

If males are scarce, the female Sierra Dome spider turns her web into a scented hankie to wave in the air. Applying an attractive scent to her web, the virgin female sends a powerful sexual signal far and wide. That's good for the female, who wants to attract many males. But what about the male who responds? His interest lies in being her only mate and in avoiding conflict. A flock of suitors does not suit him. Copulation among Sierra Dome spiders is a complicated affair lasting up to seven hours. Unlocking their complex sex organs once they've begun can take a full minute. An interrupting rival male, also attracted by the female's scented web, could spoil the mating male's whole day. So the male silences the signal by wadding up the web into a tight ball.

To see whether the male's web-wadding behavior stems from sensing a female pheromone emanating from the web or from the female herself arachnologist Paul J. Watson of Cornell University, made an extract of 84 webs of mature virgin females. He applied the love potion to an unoccupied web and introduced a male. Within a minute the male wadded up just that portion of the web Watson had sprayed. Clearly, the Sierra Dome spiders not only send and receive complex signals tuned to the circumstances, but males also interfere with female signals when it suits them.

The Waiting Game

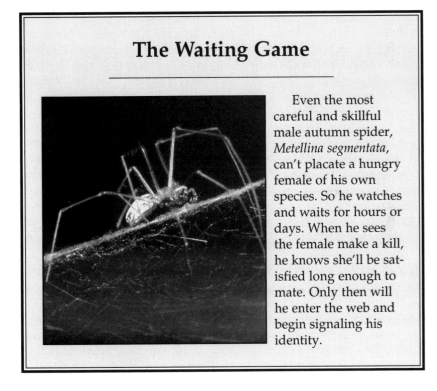

Even the most careful and skillful male autumn spider, *Metellina segmentata*, can't placate a hungry female of his own species. So he watches and waits for hours or days. When he sees the female make a kill, he knows she'll be satisfied long enough to mate. Only then will he enter the web and begin signaling his identity.

Fish

uperman's fictional heat vision notwithstanding, humans sense their world for the most part passively. We don't beam anything out of our eyes, we just respond when light hits our retinas. We do send out sound, however. Our ears, sensing this very different kind of energy, still respond passively, as minute changes in air pressure move our eardrums. But when we converse, we both send sound and receive it. We can hear ourselves and at least use this feedback to modulate our own volume and pitch. Bats emit sound and listen for variations in echoes to help them sense their surroundings. Using a sense very different from either hearing or vision, certain fish communicate and sense their surroundings with electricity.

"Just as the eyes are organs that evolution has fine-tuned and engineered to detect light optimally, ears are optimal organs for detecting sound, and taste buds are chemical receptors, these fish have very sophisticated receptor organs for detecting electric fields," reports Dr. Brian Rasnow, a California Institute of Technology physiologist.

A wide variety of fish can detect electric current. Some, like the shark, use the sense to "see" their prey. Some sharks can detect a field as weak as "five nanovolts per centimeter, which is equivalent to stretching out a one-and-a-half-volt battery over 30,000 kilometers (nearly 20,000 miles)," Rasnow observes.

But some fish have taken their electric sense a step farther; they produce electricity. The shock of the notorious electric eel has been known since before we came to understand the nature of electricity. Some smaller, gentler cousins also produce electric fields, albeit weak ones. Two families of freshwater fish have evolved the ability to create electric fields around their bodies. They can sense minute changes in these electric fields, the fields of other electric fish, the weak fields produced by all living things and disturbances in all these fields produced by inanimate objects in their environment. This sense and system of communication differs as much from vision as hearing does, Rasnow concludes.

Posing porkfish.

Flashlight fish with raised light organ just below eye.

Weakly electric fish live in muddy water and become active only at night. In their dark environment, weakly electric fish use their electric sense like many other animals use sight or hearing—to "see" where they are going, to find prey and to communicate with each other. They can distinguish the electrical discharge of their own species, and can determine the size, sex, maturity and even possibly the individual identity of any fish of their own species that passes by. Each fish, in other words, broadcasts many aspects of its identity in fluctuations and characteristics of its electric field. This sense, however, has its limits.

To appreciate these limits, take off your glasses or remove your contacts, Rasnow suggests. Then fill the room with fog. The power of an electric field fades rapidly—at twice the distance, the power falls to one-eighth—so the electric sense of these fish is limited to about half a body length. It's as if you could see a little beyond arm's length and no farther.

This short-distance communication system, however, serves the social electric fish well. In some species, social groups organize into a dominance hierarchy. The most dominant male emits the most extreme frequency, highest in some species and lowest in others. The dominant female will settle on the opposite extreme. In one species, for example, males have a frequency of about 60 Hz (the frequency of household electric current in the US that causes the buzzing you may hear in poorly grounded stereo systems), whereas females buzz at 120 Hz. Immature animals fall between these extremes.

While many animals who live in societies organized on dominance can't do anything about their status, electric fish can change their frequency. If one animal chooses to challenge the dominant male, he may begin to match the dominant male's frequency. Responding to this insult, the dominant male may emit an aggressive electrical "chirp," a sudden, brief increase in frequency—the electric-fish equivalent of a glove across the face. Once engaged in battle, two males may spend an entire night locked jaw to jaw. Finally, one male wins out, earning the exclusive right to spawn. The loser signals submission with a sudden cessation of electric discharge. In the laboratory, scientists can model an aggressive electric fish. Energizing the model provokes an attack, always to the "head" end, electrically speaking. With the electrical poles switched, the insulted male attacks the opposite end of the model.

In one species, females defend spawning territories and the dominant male will mate only with the dominant female, the one with the best spawning territory. She hangs vertically among the plants while the male emits electrical chirps 60-80 times per minute, courting all night. This electrical display stimulates the female to lay eggs, whereupon she chirps quietly, an apparent signal for the male to fertilize her eggs.

Some females try to sneak in to lay eggs among those of the dominant female. These females remain electrically quiet. If the royal couple drives them off, the intruders will lay eggs elsewhere; those eggs remain unfertilized. Nonmating females thus appear so excited by the male's electrical love song that they lay eggs even in the absence of a male. In laboratory playback experiments, scientists can duplicate this effect, playing a love song that causes lone females to lay eggs in an aquarium.

Bottom dwelling electric fish exhibit different mating behavior. A male will mate with any female who responds to his call. Successful fer-

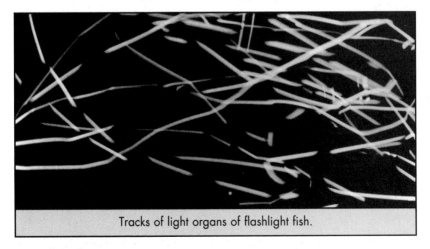

Tracks of light organs of flashlight fish.

tilization in this species is complicated compared to most fish. The female lays only one egg at a time, and the male must fertilize it just as she lays it, or it spurts to the river (or aquarium) bottom to remain unfertilized. To signal the male, the female chirps just as the egg is about to emerge.

Swimming in a muddy river and relying on your electric field can pose a problem. What happens if another electric fish swims by? The frequencies may overlap. "It's like CB radio when the channel is too noisy," Rasnow says. "The only sensible thing to do is switch to another frequency." Scientists studying electric fish call this the Jamming Avoidance Response (JAR). Two passing electric fish change frequencies, each moving its frequency slightly away from the other's. The response functions as a reflex, like your leg-jerk reflex in response to a tap on the knee. Electric fish sometimes synchronize their frequencies— for reasons scientists have yet to fathom.

Fish also communicate with sound. One species in particular caused a stir in Sausalito, California in the 1980s. A successful antipollution campaign had clarified the waters of the bay, and made the houseboats that line the shore newly fashionable. But suddenly residents began losing sleep because of a loud hum that appeared to come from the water. Some blamed a new power line. Others thought the sewage works had secretly resumed pumping at night. But the cause was, in a sense, clear water.

An elephant-nose fish.

Cleaning up the bay had produced a perfect environment for toad-fish, a slimy, bizarre-looking relative of sea robins, sculpins, and midshipmen. It turns out that toadfish hum a love song to attract mates. Males hum in chorus during breeding season, sometimes continuing for as long as an hour.

Toadfish also make two other sounds. Grunts—warnings that apparently tell rival males to back off or potential predators to stay away—last only two-tenths of a second. The so-called boat whistle that lasts nearly a second may attract females as well as humming does, and might identify individuals. In the distantly related bicolor damselfish, females can distinguish individual male chirps and males can tell the chirp of their nearest neighbor from the vocalizations of more distant males. Fish sounds function mainly in mate attraction, but are also used in school coordination, and they travel well underwater. Scientists have observed fish responding to a sound signal from half a mile away.

Fish produce sounds in a variety of ways, often with organs much less specialized for the task than the vocal apparatus of other vertebrates. Grunts appear to come from grinding teeth, the sound amplified by the air-filled swim bladder. And special muscles in or near the swim bladder itself can cause it to vibrate like a drumhead.

Cornell University neurobiologist Andrew H. Bass has studied sound production in the midshipman, sometimes called the singing fish. A set of special drumming muscles attach to the walls of the midshipman's heart-shaped swim bladder. As Bass expected, females—who don't sing—have smaller swim bladders and smaller drumming muscles. But to his surprise, Bass found that some males sport female-sized swim bladders and weak drumming muscles; they also resemble females in body size and shape. These males, who make up less than 10 percent of Bass's study population, don't sing or build nests. Their similarity to females includes their anatomy and the cellular structure of their brains. Yet they mate and pass on genes quite successfully.

These small midshipman males employ a strategy seen in many fish and in other animals as well. They are so-called satellite or sneaker males. The majority of males follow the usual route—eat and grow to a large, attractive size, compete for space, build and guard a nest, and grunt. They reap the reward of a nest full of eggs laid by several females. But sneaker males slink around the edges of territories, slip-

The brown ghost knife fish, which is weakly electric.

ping in to deposit sperm (sneak-spawning) while normal males are busy guarding and showing off. Their sperm do not fertilize many eggs in a single nest, but they can sneak-spawn in several. In a second sly strategy, some small males release their milt into the water surrounding a nest and fan the sperm-laden water toward the nest with their fins.

Fish hear—or feel—sound in two ways. Some have small bones connecting the inner ear to the swim bladder, creating in effect a single large ear. Fish have no outer ears, but they do possess an inner ear similar to those of other vertebrates.

Fish also detect vibrations in water with a unique lateral line system similar in many ways to our inner ears. The organ of hearing in our inner ears forms a coil; that of a fish lies stretched out along its side. The lateral line tube stretches the length of the fish, and sometimes branches around its head. The tube connects to the water by way of small pores in the skin and scales. Fluid fills the tube, just as in our cochlea. When a pressure wave strikes the fish, it jiggles the fluid and bends small hairs that project into the fluid in bunches. The hairs trigger nerve impulses which travel to the brain. While fish cannot determine the location of a sound detected through the single swim bladder, they can locate sounds detected by way of the lateral line.

A midshipman father in his nest.

Frogs

s dusk falls over a swampy pond, the chorus begins. First one frog croaks, a little hesitantly. Soon another joins in. Suddenly, the pond resonates with the voices of dozens of male frogs, each signaling his species, availability and qualifications as a father. Two populations of the same species, separated geographically—by a highway running through a swamp, for example—may develop dialects, slight differences in their calls. In most species, only males call, although the female midwife toad outshouts her mate. A sudden foreign sound silences them one and all. But soon a single voice starts again, followed by another and another.

The sound of a pond full of frogs can reach deafening levels, far out of proportion to the tiny bodies producing the sound. An air sac on the floor of the frog's mouth enables it to do two remarkable things. First, when it's expanded, the sac acts as a resonator, like the hollow body of a violin. Second, by forcing air into the sac from the lungs, then back into the lungs, a frog can croak continuously, even underwater. The loudest frogs breed in temporary ponds. When the water is available, males muster mates quickly. Frogs lay eggs in water only, and tadpoles must grow legs before temporary ponds dry up.

For sheer cacophony, the year-round singing of tropical frogs beats all. In the tropical rain forest, one of the richest ecosystems on earth, a swamp houses many frog species. Males would gain nothing by attracting and attempting to mate with females of another species, so they have to make their signals cut through the din.

The male *Leptodactylus ocellatus*, which lives in South America, calls at about 250 to 500 Hz (roughly an octave, from middle C to C above middle C). His volume is no match, however, for a neighboring species, which overlaps his frequency range and can croak 40 decibels louder. That's like the difference between a conversational tone and the noise on a factory floor. *L. ocellatus*, however, has evolved a way to circumvent this potential problem—by croaking underwater.

> With a fully inflated air sac, this frog emits a sound proportionally much bigger than itself.

Carpenter frog.

Because sound travels well underwater, but doesn't cross the interface into the air, *L. ocellatus* avoids sonic competition without changing his preferred octave. In a rich environment, species divide the environment up into tinier and tinier slices, each becoming a specialist.

Singing underwater represents an extreme method of dividing the "airwaves." Most frogs croak distinctively by varying the pitch of their calls, or by varying the pattern of croaks in a song. Even when they can distinguish the call of males of their own species, female frogs must determine the direction from which it comes. This presents a problem.

Humans and other mammals can determine the direction of a sound because their brains detect differences in loudness and time of arrival at each of their two ears. For this system to work, the wavelength of the sound must be much smaller than the distance between the two ears. The longer the wavelength, the harder it is to tell where the sound is coming from. The crest of a long sound wave may hit the near ear only a tiny fraction of a second before it hits the far ear. A comparatively long time later, the next crest reaches the ears. This is why some stereo systems contain only a single low-frequency woofer, which the listener can place anywhere in the room. By contrast, the high-frequency tweeters and midrange speakers must be separated and properly placed to give the stereo illusion of an orchestra in the room.

Humans hear only up to 20,000 Hz, and the highest note on the piano is only a little above 4,000 Hz. But many mammals hear extremely high frequencies, up to 60,000 Hz, quite well. These sound waves are much shorter than even the tiniest mammal heads. Frogs' high-frequency hearing is not as sharp. It tops out at 10,000 Hz, making the distance between their eardrums too short to localize the highest pitches they hear. So how do females find their mates? Scientists studying the *coqui* frog of Puerto Rico, *Eleutherodactylus coqui*, think they have a lead.

One way frogs can localize sound is by hearing a sound twice in the same ear. Because of the frog's head anatomy, sound can travel from the middle ear down the eustachian tube, across the frog's mouth cavity, up the other ear's eustachian tube and into the middle ear. Sound can then affect the near ear of a frog, for example, twice. First when the sound strikes the eardrum, and again when it arrives a bit later by way of the mouth and eustachian tubes from the far ear. These two sounds may arrive out of phase, with peaks and troughs interacting in a way that gives the frog's brain a clue as to where the sound came from. This pathway doesn't affect mammals because our mouth and ear anatomy differs from that of frogs.

By mapping vibration in the frog's body with lasers, scientists have found that a particular spot over the lungs of some species vibrates along with the eardrum. Researchers propose that the body wall here acts like a large eardrum, providing yet another pathway for sound to follow to the eardrum—from lung to mouth to eustachian tube to middle ear. This mechanism may further increase the precision with which female frogs find their princes.

A female frog targets the territory of the wonderful voice she hears. But when she finally sees the male frog, she may get a ringer. Some male

American toads in amplexus, the typical clasping mating embrace.

Two poison-arrow frogs.

frogs don't call at all. Instead, they silently invade a strong caller's territory, hoping to intercept a female as she approaches. For all the female knows, she's found the owner of both the beautiful voice and the territory. Such so-called satellite males may not mate often, but the strategy at least gives males that can't call competitively a fighting chance to pass on some genes.

Satellite mating works for males who don't meet the usual strict standards of females. But what is a female frog's best strategy? Mac F. Given, a biologist at Neumann College in Aston, Pennsylvania, found that, unlike most frogs and toads, female carpenter frogs have their own call, and he suspects they use these calls as a way of double-checking before they mate.

Male carpenter frogs sing a short song of up to ten notes to advertise their territory and availability, and they continue to call and defend their territory for up to three months. Another male may respond to this long call by a single-note chirp that Given calls the aggressive call, triggering a train of aggressive calls and responses until the territory owner grapples with the intruding male. The intruder usually gives in, chirping a release call. He then splashes on his way.

A female carpenter frog responds to the male song just like a male, inciting the same seek-and-fight behavior in the male. But she never gives the release call, and the "wrestling" match turns into amplexus, during which the male clings to the female's back in position to fertilize the eggs she lays. Considering the risk the female runs in possibly attracting a predator, Given wondered what the female gains. He thinks she may call to stymie satellite males. A satellite male dare not return her chirp for fear of attracting the territorial male. So by calling back and forth, the female makes sure she mates with the real prince, not a sly, silent impostor.

Bullfrogs.

Birds

Learning to Sing

morning in spring would not be complete without bird-song, ranging from long and complex, such as the winter wren's high-pitched, multi-note melody, to the Townsend's solitaire's single plaintive whistle. To study song learning, researchers isolate maturing male birds from all singing males. In classic studies, English ethologist W.H. Thorpe raised European chaffinches in complete isolation. These birds sang only the skeleton of a chaffinch song when they matured the following spring. In contrast, a chaffinch exposed to adult male "tutors" singing during the first few weeks of life grew up to sing several fleshed-out variations on the basic song. He presumably could compare his simple first attempts with a model remembered from the previous summer.

Raising chicks in complete isolation, however, can produce maladjusted birds, according to Meredith J. West of the University of North Carolina and Andrew P. King, of Duke University. To avoid this problem, they raised hatchling male cowbirds from North Carolina with adult females from Texas. To their surprise, the males grew up singing with a distinct Texas twang. But how had the silent females "taught" the dialect?

By careful observation of cowbirds in the wild, West and King solved the puzzle. Males sing four to seven variations built on a basic gurgle, gurgle whistle theme. Each male sings a unique set of variations. When a female hears one she likes, she signals her approval by flipping one wing up and out, batting her eyelashes as it were. The behavior represents a rare example of an expression of approval among non primate animals. The signal takes place in the blink of an eye, only a few thousandths of a second, during the male's one-second song. But that's enough for the males. Once they know what a female likes, they concentrate on just that variation. The researchers draw parallels with humans. Babies show approval of adult vocalizations with smiles and alert looks. Their caretakers learn by repeating the

Red-headed woodpeckers, by Audubon.

kind of vocalizations—soothing baby talk, for example—that triggered the approval. As babies grow, however, their preferences change. By the age of a year and a half, they prefer a more adult-sounding vocal style, and the easy-to-train adults comply.

The "dialects" of birds probably have no more significance than those of humans, comments Michael D. Beecher of the University of Washington. "Birds learn songs, presumably, because of benefits to the individual, and these benefits appear to relate to learning songs from (and thus sharing song types with) your immediate neighbors (following

A Siberian songbird.

dispersal from the nest and natal area). If you learn songs from your immediate neighbors, one larger-scale consequence is area dialects." Matching songs with your neighbors, he says, may play a crucial role in the social structure of a species.

Bird songs send two related signals, as in most male animal vocalizations; first, to announce the male's breeding status to females, and second, to post his territory with auditory "Keep Out" signs. Song sparrows sing eight or nine distinct songs, several of which they share with close neighbors. And they're not alone. About 70 percent of all songbird species sing more than one song type. Evolutionary explanations of these repertoires are a topic of debate among bird behaviorists, many seeing the song variants as interchangeable. Beecher, however, sees a more complex communication going on. A young male learns songs in his repertoire from three or four older males. Then when he sets up and begins to defend his own territory, he can reply to songs of neighbors—both his former tutors and his young classmates—because he shares song variants with them.

Hearing a neighbor's song from near or inside his territory, a male song sparrow could defend his territory employing various strategies. He could repeat the same song type or reply with a different song variation—one both birds share. Song sparrows Beecher has studied typically share about 40 percent of their song types, and rarely share no common song types. In a field study, Beecher played recorded songs to defending song sparrows. Sparrows that heard a neighbor's song

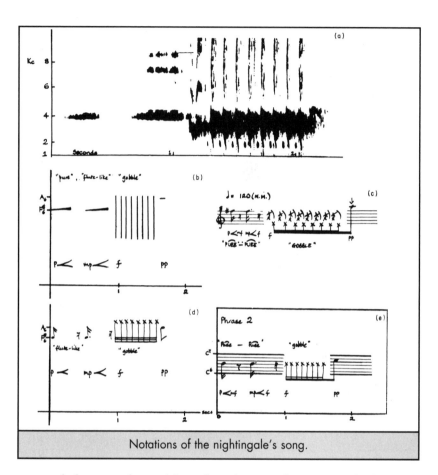

Notations of the nightingale's song.

responded most often with a shared song—but not with the song they'd just heard. Presented with a strange song or their own song, they rarely used a shared song to respond. Instead, they matched the song they had just heard. Beecher suggests that using a song from a shared repertoire helps to maintain a status quo. It's a reply of low intensity and represents a mild challenge, more like a hard stare than a shouted, "Get lost!" Matching a song more precisely represents a direct challenge, which can escalate into an energy-sapping chase.

Most birds, like Thorpe's chaffinches, learn their songs during a critical period of growth. Chaffinches learn songs before they mature sexually. Similarly, white-crowned sparrows of the American West learn songs during their first summer of life. After that they do not learn songs. Humans also learn language best at a young age. After our teens, few of us learn to speak a foreign language without a distinct accent. A final parallel lies in the genetic component of learning communication. Experiments with songbirds show they can learn variations on their species-specific songs—even variations never seen in nature, such as songs with the first and second halves switched. But they can't learn songs of other species. (Of course, some species specialize in imitating other species' songs.) As humans, we seem genetically programmed to learn human language. We sponge up language. But we don't learn to imitate other sounds surrounding us as we grow, even if raised in isolation from human speech.

Barnyard Chickens and Junglefowl

Consider the difference between two alarm calls: "Hey!" and "Duck!" The first could cause the receiver to just look around and, unaware, move into danger rather than out of it. The second conveys information, not just alarm. For years, researchers considered all animal alarm calls devoid of meaning, like the inarticulate human shout. But careful observation in the field uncovered new insights. Animals do not always respond the same way to danger. Even the lowly barnyard bantam possesses a small vocabulary of danger signals. In response to the sight of a weasel invading the yard, a bantam emits a high-pitched "Kuk Kuk Kuk." But if she spies a hawk circling overhead, she'll shriek a single long call.

To mark a behavior as deliberate communication, scientists look for some indication that the sender *intends* to communicate. Researchers also look for a response from the receiver. In their alarm calls, bantams satisfy both criteria.

Bantams issue calls alerting other birds about predators much less often if no other bantams are around. Known as the "audience effect," this phenomenon seems to indicate awareness of whether or not the call will do any good. If it were merely an emotional alarm, the chickens would squawk under all circumstances, just as a human might scream at a scary sight, whether or not anyone else is around. What's more, bantams have two calls that trigger two clearly different responses. A series of clucks indicates a ground predator and a scream a flying predator. When a bantam hears clucks, she cranes her neck, scans the ground, and runs to the middle of the yard, where she might see a weasel creeping from under the henhouse. A bantam who hears a scream looks upward and runs for cover. Fortunately for researchers, bantam hens and roosters see and believe in television. A video of a racoon or a red-tailed hawk elicits the appropriate call, and the call elicits the appropriate response.

Both bantams and Burmese Red junglefowl, the wild ancestors of domestic chickens, also call to announce their discovery of food. The call, a low series of single-note clucks called tidbitting, conveys more than mere satisfaction at finding food. Cocks rarely tidbit except on discovering food and in sight of a hen. Hens rarely cluck about food unless their chicks are nearby. Interestingly, junglefowl tune their food call, depending on the food they find. For an excellent food source—say a juicy grub—they cluck more often and more rapidly. Hens respond less readily to "low-quality" food calls. Although newly hatched chicks respond to high-quality and low-quality food calls differently, they also learn to discriminate better as they gain experience.

On occasion, domestic cocks will emit a food call when, in fact, they have not found any food. They may also advertise high-quality food when the food discovered is actually not that good. Some researchers suggest that such behavior is deceptive. But since the call functions both as a food alert and as a mating call, the cock may not be guilty of false advertising.

Bowerbirds

With its glossy dark purple coat, striking pale blue eyes, and yellow beak, the male satin bowerbird is not plain, yet its plumage nowhere near matches the fantastic feathers of the closely related birds of paradise. Instead of displaying gaudy feathers, bowerbirds as a group employ a remarkable method to catch the eye of their brides-to-be. They build trysting places, called bowers, unlike the structures of any other animal. Using the bowers as a stage, the birds show off their dancing and singing prowess to admiring females.

Of the 18 bowerbird species, 14 build bowers, ranging from simple to complex. The toothbilled bowerbird clears a patch of forest floor and covers it with painstakingly collected fresh leaves. He lays each leaf with its pale underside uppermost, giving the carpet a light shining appearance in the forest. At the other extreme, the striped gardener bowerbird meticulously clears the base of a sapling and the ground around it. He then weaves a layer of twigs around the sapling. This structure, called a Maypole, serves as a centerpiece for his mating dance. He covers the dance floor with moss and decorations, such as shells, feathers, and bright leaves. He may even find buttons, bottle caps and other bric-a-brac, all color-matched to his natural trinkets. Finally, the gardener bowerbird tops his dance floor with a domed structure of twigs three feet tall and nearly twice that in diameter, then dances around the Maypole inside.

A bowerbird exploring a bower.

The best-studied of all bowerbirds, the satin, builds a bower between these two extremes. It consists of two parallel fences that look like vertically flattened haystacks. The promenade between the fences faces the noonday sun (north in Australia) and opens on a dancing platform covered with bright yellow straw. The male satin bowerbird decorates his stage with prized blue parrot feathers, blue and yellow flowers, the shiny wings of cicadas, and other shiny items he can find. He delineates the boundaries of the stage with larger objects, including land snail shells. What's more, he maintains a supply of small objects near the bower to hold in his beak while he dances. The shiny blue objects on the bright yellow background provide a glowing contrast in the dark forest. He enhances this effect by pruning nearby trees to let more sunlight shine through—all to attract the attention of a female.

University of Maryland zoologist Gerald Borgia systematically studied satin bowerbirds, partly to determine how female selection might operate in this species. Female satin bowerbirds examine several bowers in a small territory, using them to assess the qualities of the males maintaining them. Borgia found that building a bower well takes some experience. Older males build neater, sturdier bowers and accumulate a greater treasure of blue objects, especially rare blue feathers. They also give more "refined" courtship calls. Foraging for blue feathers is no easy task. Because they are extremely rare, most birds steal them from other bowers. The marauding birds may also trash the bower as they invade to steal trinkets. Protecting and maintaining a bower requires more energy and skill than building it in the first place.

Borgia recorded bowerbird behavior on film, and also manipulated their habitat. Systematically, he removed ornaments from the bowers of the more successful birds. Ultimately, these males attracted fewer mates. Borgia also introduced numbered blue feathers into the habitat and found that they ended up in the bowers of the most successful males. Older, more experienced satin bowerbirds may mate with up to 33 females in a season. Each female mates only once, nests in a tree nearby, and incubates the eggs alone. Less successful males may not mate at all.

Alex Speaks

Humans have been fascinated with birdsongs probably since our ancestors first evolved the capacity to be fascinated. Early on, listening to and appreciating birds paid off in food—the bird itself, its eggs or its food. Later, birdsong became to humans a natural art form that we still highly prize.

But nothing beats hearing the sound of your own words. Since the time of Aristotle, people have known that parrots can talk back. Several species can mimic human speech, from "Hello" to "Polly want a cracker."

But does imitation imply understanding? Irene Pepperberg of the University of Arizona set out in the spring of 1977 to explore the ability of parrots to mimic human speech as a tool for understanding what they may have to say. She chose a champion speaking species, the African Gray parrot, in particular an African Gray named Alex, who knows more than 100 words.

Alex, Pepperberg found, can do much more than learn a hundred words, including 90 names of objects. He can also answer questions. Pepperberg asks "What's this?" (holding up a green key). "Green key," Alex responds. Holding up a blue pentagonal wooden block, Pepperberg asks "What color?" and Alex answers "Blue wood." But if she instead asks, "What shape?" Alex knows to respond "Five-corner wood." (Apparently "pentagonal" is a bit beyond Alex.) Alex can also pick a red key out of a collection that includes red objects that are not keys and keys that are not red.

Birds, it turns out, can keep track of counting things quite well. Scientists believe that crows keep track of both the number of caws they hear and their rhythm to identify each other. Pepperberg likens the birds' ability to how we sing "Deck the Halls" at Christmas without consciously counting the number of fa-la-las. Timing is a critical component of singing for songbirds, involving the number, sequence, and duration of notes.

Pepperberg set out a bewildering (to us) collection of objects—wooden tongue depressors, spools, and metal, paper, and plastic cups among other things. Each object fell into a color category (rose, green, purple, blue, yellow, orange and gray). Pepperberg and her students then asked Alex, "How many blue cups?" or "How many orange keys?" In each test, experimenters arranged the objects to test Alex's ability to count a particular number, starting with one object and advancing to six. Alex answered more than 83 percent of the questions correctly. He made about the same number of mistakes no matter how many objects he was asked to count. Unless Alex's ability to visually segregate objects that met the criteria and then perceive their total without counting greatly exceeds ours, Pepperberg concludes, Alex counts in the same sense we do.

Pepperberg cautiously draws some parallels between the way Alex learned and the way preverbal children learn. Young children talk to themselves when they're in the company of adults and other children. They also talk to themselves when they are alone, often practicing new words they have just heard. Alex also practiced when left alone. Pepperberg recorded Alex's vocalizations each evening after his trainers left. In these monologues, Alex practiced words his trainers had been teaching during the day—words he had never spoken before. Alex also often practiced words to himself before using them in reply to trainers. He always practiced privately before words entered his vocabulary reliably.

Cetaceans

Dolphins

Among animals, birds take first place in imitating sounds, including human words. But perhaps surprisingly, marine mammals place a respectable second. Certain harbor seals have learned to bark out (barely) recognizable words, and whales imitate each other during the development of long, complex songs. But the most widely studied marine mammal vocalizers are dolphins.

Dolphins communicate using a great number of different sounds, from repetitive clicks for echolocation (and possibly for communication) to whistles and grunts. In captivity, they can even imitate some human words. Their echolocation appears truly remarkable. A blindfolded dolphin can find an object the size of a penny at the bottom of a swimming pool and can distinguish small objects by their shape and their composition. But most remarkable is the way dolphins vocalize to communicate.

At first, the whistle sounds dolphins make all appear to say, "Hi! Hi!" But research conducted since the 1960s has suggested that dolphins each generate a unique whistle, known as a signature whistle, implying that dolphins produce many different kinds of whistles—at least 10 to 25, the number of individuals in an average group. Furthermore, they must learn the signature whistles of every other dolphin in the group. When scientists listened to whistles of more than one hundred dolphins, they concluded that dolphins do not choose a signature from a fixed set of whistles, but develop their own individual whistle. As the dolphin matures, its signature becomes somewhat stereotyped. To identify individuals by their whistles, dolphins must be able to link an individual whistle with an individual dolphin.

Behaviorists Peter Tyack of Stanford University and Laela Sayigh of the University of North Carolina at Wilmington studied wild dolphins near Sarasota, Florida. These dolphins, a group studied for decades by Randal Wells of the Chicago Zoological Society, remain in one area, so researchers can repeatedly find individuals which they recognize by the animals' individual markings. Because researchers cannot tell which dolphin is making a particu-

Bottlenose dolphins.

"Hi! I'm not just telling you my name."

lar sound, they record the dolphins individually after corralling them with a net. While the technique reproduces only a seminatural environment, nonetheless, it has yielded intriguing information.

First, Sayigh confirmed that individual dolphins have recognizable signature whistles and that those whistles remain stable for at least a decade. She also found that mothers and their calves remain in vocal contact when one is in the corral and the other swims nearby. When Sayigh began recording newborn calf whistles, she heard a faint, quavery sound that varied from whistle to whistle, like the written signature of a young child. By the age of one, the calf had firmed up an individual whistle that then remained more or less constant.

Sayigh also discovered that male and female calves develop whistles differently. A female calf learns a whistle that is distinctly different from her mother's. A brother's whistle, on the other hand, develops from a variable baby whistle to an adult form that closely resembles his mother's. In this population of dolphins, at least, mothers and their newborn calves move together with their grandmothers and other females, forming a long-lasting group. If young females, their mothers, grandmothers, and other females emitted similar whistles, it could lead to confusion—as in a human family with two members who have the same name: the wrong person is always taking the telephone call. Male calves leave the group when they mature, so they stand little chance of being misidentified.

While the modification of signature calls appears to rely on learning, Sayigh has not ruled out a genetic basis. For example, if the signature call were genetic, it might reside on the X chromosome. A female would inherit two X chromosomes, one from her father and one from her mother. Her signature would then differ from both. A male, on the other hand, inherits only one X, from his mother, and his call would resemble hers.

Studying signature whistles poses problems. Although we can hear dolphin whistles under water, we cannot locate the source. Nor can we locate the source when the whistles are picked up by underwater microphones and broadcast topside. Isolating the dolphins solves the problem of identifying which one is squeaking. But when two dolphins communicate by way of a sort of underwater walkie-talkie—as they did in one experiment—scientists found that isolation severely disrupts their social communication.

Peter Tyack invented a solution which allows him to record whistles of two or more individual dolphins reliably. His invention, called the vocalight, is a three-and-a-half-inch-long device that looks like a cross between a small flashlight and a model of *Star Trek*'s U.S.S. *Enterprise*. A suction cup holds the device to the dolphin's slick skin in front of the blowhole. A row of light-emitting diodes points to the front and a battery case to the back. Inside, a microphone picks up the dolphin's sounds and electronics illuminate the diodes in response to the sound. The louder the sound, the more diodes light up.

Tyack studied Scotty and Spray, dolphins that lived in the Sealand aquarium in Brewster, Massachusetts. One dolphin was given red diodes and the other green. Whenever they heard a whistle picked up by an underwater microphone, Tyack and others called out the color and number of diodes they observed. A tape recorder captured the whistles and observers' reports.

By displaying them as sonograms, Tyack found more than three quarters of the whistles fell into two easy-to-recognize categories. Type 1 rises, swoops down to a lower pitched half-second-long whistle, then rises slightly. Type 2 also rises, falls and levels out, but then rises abruptly at the end. Although both dolphins produced both kinds of whistles, Tyack noticed individual variations. Spray's Type 1 was not identical to Scotty's Type 1, and their Type 2 whistles differed even more markedly. Interestingly, the dolphins split their whistles between the two types differently. Two thirds of Spray's whistles fell into Type 1, while almost three quarters of Scotty's were Type 2. Tyack interprets these results to mean that Type 1 was Spray's signature and Type 2 Scotty's. Each dolphin imitated the other's signature whistle some of the time, Tyack contends, perhaps using the other dolphin's signature whistle as a label or name.

The dolphins also whistled variations on their two themes—they left parts out, varied duration, changed pitch slightly, and—as seen on a sonogram—varied the shape of the sounds. Tyack maintains the dolphins could detect these variations and many whistles that fell into neither category.

Unfortunately Spray died, leaving Scotty in isolation. Tyack returned two years later, curious to see if Scotty's vocalizations had changed. Scotty no longer used Spray's favorite whistle. And his whistles in general had become quieter and briefer, lending weight to the idea that dolphin whistles function as social communication.

Seen from this perspective, it appears that dolphin vocabulary may well encompass much more than "Hi! I'm Scotty" and "Hi! I'm Spray."

Although dolphin vocalization in the wild may go beyond mere signature whistles, earlier attempts by scientists to read imitations of human sounds into their squeaky voices are not well-founded (see sidebar). Even today, it's difficult to rapidly evaluate their high-pitched sounds. Furthermore, their sleek, fingerless flippers make it awkward for dolphins to produce the kinds of artificial communication—hand gestures and keyboard use—that apes can perform.

Still, research has moved forward on dolphins' ability to understand language. Louis M. Herman and his colleagues at the Kewalo Basin Marine Mammal Laboratory at the University of Hawaii at Manoa have taught four wild dolphins artificial languages. One language consists of computer-generated, high-pitched "words." In the second, a trainer conveys words with hand and arm gestures. Neither language bears any resemblance to human language, except in that both contain grammatical rules. Each language comprises about 40 words: nouns such as "channel," "gate," "person" and "ball"; verbs such as "under" (the "go" implied) and "fetch"; and modifiers such as "surface," "bottom," "right" and "left."

Using these languages, Herman can ask a number of questions about dolphins' ability to comprehend words as referents to specific objects and to understand that the order of the words makes a difference. Herman's dolphins never received grammar lessons. Instead, they learned the rules by example. In the high-pitched vocal language learned by Phoenix, "Phoenix Ake under," for example, means the trainer wants Phoenix to swim under Akeakamai (Ake for short). The gestural language learned by Ake included a reverse grammar, to prevent word-by-word responses. The dolphin receiving a gestural command must understand the full two- or three-word sentence before responding.

Herman found that the dolphins learned both what each word referred to and how to interpret the order of words. For example, Ake distinguished between gestural sentences such as, "right hoop left Frisbee fetch," and "left hoop right Frisbee fetch." The first sentence means "Take the Frisbee on your left to the hoop on your right." Success with such sentences demonstrates dolphins ability to understand both the semantic (word meaning) and syntactic (sentence pattern) components of the language, Herman writes.

Ake, however, has gone beyond merely responding as trained. She even invents her own logical responses to unusual situations. If a trainer commands, "Frisbee hoop in," a two-gesture sequence asking Ake to put the hoop on top of the Frisbee, Ake normally complies. But Ake has also learned to press large paddle-shaped switches labeled "yes" and "no." For example, if neither the Frisbee nor the hoop were available, the trainers would expect her to press the "no" paddle. Sometimes, when both objects were in the tank, Ake would put the hoop—the object the command asked her to manipulate—on the "yes" paddle, a behavior she invented. When the Frisbee is not in the tank, Ake will put the hoop on the "no" paddle. When the hoop is missing, she will not move the Frisbee—the trainer didn't ask her to—but will press the "no" paddle, meaning she can't comply with the command.

Ake responds to novel combinations of words correctly, showing that she understands both the words and the sequence. But what does she do with a construction that makes no sense? In that case, Ake will move an object to make it possible to comply with a command. She will lift a hoop that's lying on the bottom of the tank so she can swim through it, for example. If, instead, she sees a sentence that gives an impossible command, such as "person water fetch," Ake does nothing. She can't move the water to the person. A more complex kind of "impossible" command, "person water hoop fetch," contains too many nouns. In this instance, Ake sensibly ignores the noun water and interprets the sentence as "take the hoop to the person." She has interpreted the impossible "water" as a mistake, much like we might when a duplicated word represents a typographical error. Herman sees these untaught responses as indications that dolphins make a mental representation of the grammatical rules of their artificial language. By referring to their memory of the structure, they can make sense of novel, or even nonsensical, sentences.

A picture is worth a thousand words, the saying goes. But in our digital age, is a picture composed of every other pixel worth merely five hundred? On the contrary, we can still "read" a picture with a large part of digital detail missing, an image of Abraham Lincoln made of only a few dozen blocks of color, for example. Herman wondered whether Phoenix and Ake could "read" gestural communication with limited detail. First, on an underwater TV monitor, he presented a videotape of a trainer to the dolphins. The dolphins responded almost as well as to a live trainer. Next, Herman blanked out the head and torso on the screen, leaving the arms only. Herman then removed the arms from the screen image and, finally, he showed only two moving white spots to represent human hands. The dolphins responded to all of these presentations. Correct responses fell off only with the last, most abstract presentation. Even this minimal communication evoked more correct responses than chance would predict. College students with four months of experience in the gestural language used in these experiments responded about as well as the dolphins.

From these and other studies, Herman concludes that dolphins use words of the artificial languages to refer to objects in an abstract way, and can make sense of artificial grammar as well.

Baleen Whales

A scientist from outer space scanning the earth for life forms might exclaim, "There be whales here." For the most powerful voice on our planet belongs to the blue whale—the largest animal ever to have lived on our planet. Comparable in sonic energy to the twin booster rockets of the space shuttle, the blue whale's moan can cross an ocean and echo from the other side.

Such long-distance calls also represent the first infrasound of biological origin accurately described. Nonbiological sources of infrasound abound—including thunder, air turbulence, jet engines, volcanoes, earthquakes on land, and waves and ships in the ocean. In the early 1950s, with new electronic underwater listening devices spun-off from World War II research, scientists recorded infrasonic pulses of sound. Instrumentation recorded second-long tones at 20 Hz, about the lowest bass humans can hear. First thought to be Russian submarines, then a

Breaching humpback whale.

whale's heartbeat, scientists finally realized the sounds were the vocalizations of fin whales.

Fin and blue whales belong to the order Cetacea. The order includes the toothed whales, such as killer and sperm whales, as well as dolphins; and the baleen whales, such as the right, fin, blue and humpback. Baleen whales eat small crustaceans, filtering them from huge mouthfuls of water with a plate called baleen, a fingernail-like material that gives this group its name.

Their mode of eating is not the only difference between toothed and baleen whales. The eating habits of baleen require them to migrate over long distances. The humpback, for example, mates in Hawaiian waters and feeds—only in the summer—near Alaska. Baleen whales also differ in growth and social structure, both of which affect their means of communication. Besides being the biggest animal, blue whales mature the fastest. A blue whale calf grows by 175 pounds and more than an inch a day. The calf only nurses for six months—at which time she's at least half her adult size—and sexually matures in less than five years.

The relationship between a mother and her calf, however short, is the longest-lasting one in baleen whale society. Although groups migrate, feed, and breed together, they remain fluid, and a particular group structure may last only hours. No well-studied species of baleen whale forms monogamous mating pairs, so when males find females, the competition heats up.

Male whales, like males of many other species, appear to advertise themselves vocally. But no other animals do so in such an elaborate way. Humpback whales sing complex songs, which can last up to 20 minutes. The songs comprise several parts, each of many notes, ranging from a bass rumble to a shrill squeak. Each male sings a song repeatedly, part for part, note for note, sometimes spending hours in a single concert.

While dolphins identify themselves and each other with signature whistles, scientists have yet to learn whether baleen whale songs identify individuals. Whales within a group all sing slight variations of a single song. While one could conclude that the song is genetically hardwired, the song changes over time. Gradually throughout the singing season, note by note, the song evolves, with the result that today's community song may bear little resemblance to last year's. The transformation does not come about by a complete change in the makeup of the group—researchers know this because individual whales can be identified by markings from year to year.

What are whales doing with their songs and infrasonic calls? No one knows for sure. The songs appear important in mating, and infrasonic calls may allow whales to keep in touch over long distances. One scientist estimates that a fin whale can detect a broadcast by another fin 3,000 miles away. Infrasonic calls travel such distances intact under water because the sound waves not only bounce off the under surface of the water, but echo off boundaries between cold and warm water. Instead of spreading, as sound does in air, infrasound travels in a corridor, dissipating much less quickly. What's more, the wavelength of a particular pitch increases more than four times in water, making it more effective as a long-distance signal.

Where Does the Dolphin Produce Sound?

Despite decades of research on the sounds made by dolphins, the precise source of their sound remains unknown. Now scientists using computed tomographs (CT scans) have homed in on the source of the high-pitched clicks dolphins use in echolocation.

A biologist and an acoustical physicist from the University of California at Santa Cruz have teamed up to create a two-dimensional computer model of a dolphin skull, showing the hollow air sacs sur-

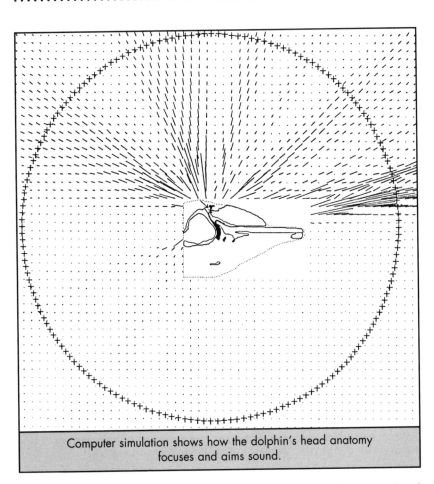

Computer simulation shows how the dolphin's head anatomy
focuses and aims sound.

rounding the internal parts of the blowhole and the fatty forehead
structure called a melon. Physicist James L. Aroyan, marine biologist
Ted W. Cranford, and their colleagues used a supercomputer to simu-
late the effect these structures—one by one and in combination—produced on a simulated dolphin click emanating from three spots within
the dolphin's head.

Using their model, the scientists moved the source of the sound, and
recorded the pattern of sound emanating from the head. The skull by
itself focused the sound to some extent upward and forward. Adding

John Lilly: Taking Fascination Too Far for Science

John Lilly, a physician and neurophysiologist, became fascinated with dolphins in the early 1960s. He studied their brains, trying to understand how their brains controlled their complex sounds. By chance, Lilly says, he once heard a dolphin imitate some of the sounds in a conversation it had overheard. Lilly was hooked. He began to study the ability of dolphins to imitate human speech, slowing down tapes of dolphins and claiming that he heard clear repetitions of human words and phrases. Only some observers could understand words in these vocalizations, and even among those, interpretations varied.

Lilly also found the relative size and anatomical complexity of the dolphin brain remarkable. He proposed and defended the idea

air sacs, which include a tissue-air interface particularly effective at reflecting sound, focused the resulting click even more toward the front. The skull and air sacs appear to act as an acoustical mirror, reflecting the sound in much the same way the curved mirror in a flashlight reflects light into a condensed, directional beam. The melon fine-focuses the sound into a more intense beam aimed forward and slightly upward from the axis of the dolphin's beak. This angle corresponds well with those measured in live clicking dolphins.

By moving the source of sound around, Aroyan and Cranford found the most likely source of dolphin clicks to be structures the scientists call "monkey lips." Cranford had previously peered down a clicking dolphin's blowhole with an endoscope—like those used to investigate

that the brain size of the dolphins indicated that they possessed language as rich as ours and went on to claim that it might be possible for us to carry on intellectual and philosophical discussions with them, if we could only learn to speak a common tongue. Lilly reported that "[Human] knowledge of cetacean intelligence and computational abilities, and of the necessities for survival in the sea are primitive and incomplete." But he also claimed that "The Cetacea are sensitive, compassionate, ethical, philosophical, and have ancient 'vocal' histories that their young must learn."

Most scientists concluded that Lilly's speculations were extremely unlikely and completely unsupported.

The great popular success of Lilly's 1961 book, *Man and Dolphin*, contributed to the widely held belief that dolphins are especially intelligent and perhaps that animals in general have languages akin to human language. While Lilly's work may have increased public awareness of the need to protect and conserve dolphins, it did so at the cost of public misunderstanding. Dolphins—and for that matter all animals—need not be like humans to deserve our respect.

human throats. Those observations, combined with CT scans of a dolphin that died of natural causes, allowed Cranford to propose a pair of fingernail-like structures inside the blowhole—the monkey lips—as the source of sound. When Aroyan simulated several other locations for the source of sound, the focus and directionality of the resulting sound beam worsened.

To produce a sound, a dolphin exhales high-pressure air through its nasal passages to its blowhole. On the way, the air must pass between the monkey lips, like air blown hard between two pieces of grass, or an oboist's breath between the two reeds of his instrument. The sound echoes off the skull and air sacs and passes through the melon into the water as a tightly focused beam of sound energy.

The Contrabasses

eat shimmers in the air, distorting the tan grass and gray-green, scrubby trees near a family of African elephants. They graze languidly, catching what shade they can from the sparse trees. Suddenly, they all lift their heads in unison, flop their big ears forward and begin to march away, as if alerted by an inaudible air-raid siren. Miles away, they blend with another group.

A bull in musth—physiologically ready to mate and searching for a female—mysteriously avoids other males, but marches miles directly to a female in heat. Old Africa hands used to call both these phenomena "elephant ESP."

A scientist studying the movements of elephants fitted with radio-tracking collars documents the odd coordination between families of cows and calves. He repeatedly tracks two separate groups moving in unison, for hours, days, and even weeks at a time. They turn together, maintaining parallel tracks miles apart. Sometimes, the groups simultaneously change direction, moving directly toward each other and blending. While the elephants may use their keen sense of smell to coordinate movement, wind often carries odors in the wrong direction, so the scientist concludes that odor alone cannot account for the coordinated movements of elephants.

Several bulls dip their dusty trunks into a water hole, in Namibia's Etosha National Park, savoring the stark contrast to the parched air they breathe. Suddenly, two look up, spread their ears wide, and crunch more than half a mile through the brush to find, not a female in estrus, but a pair of biologists and a Volkswagen van with a huge speaker mounted on top. The elephants, possibly taken aback, march on past. The biologists, Loki Osborn and Russel A. Charif of the Bioacoustics Research Program at Cornell University, watch with relief. They had broadcast a call recorded from a female in estrus, but neither they nor the rest of their team, videotaping in a tower near the water hole, heard a thing. The sound, below the threshold of human hearing, forms part of the remarkable infrasonic elephant communication system.

African elephant family at waterhole.

Elephant ablutions.

We humans can hear many elephant calls—from the famous shrill trumpets to low groans. But until Katherine B. Payne of Cornell analyzed a tape she'd made of Asian elephants at Portland, Oregon's Washington Park Zoo, no one knew that the deepest elephant sounds we hear, called grunts or rumbles, were merely the mild overtones of sounds so low and powerful they travel unhampered for miles through Asian forest. African elephants use similar signals.

Elephants live in complex societies and, like any social animal, they must communicate. These largest of land animals communicate with every sense—touch, taste, smell, vision and hearing. All work at close range, within a small band of elephants browsing together, or between mother and calf, or mating male and female, for example. With their long trunks, elephants can keep track of odors on the ground as they walk head up and they routinely touch and smell each others' bodies with their trunks.

But it was their sense of hearing that baffled early naturalists and makes long-distance communication—and therefore elephant society and mating—possible. Small groups of related adult females and their young of both sexes form the basic unit in elephant society, called a family. Females remain in families for life; the family often consists of three generations and may remain stable for decades or even centuries. Families associate with one to five other families, probably consisting of more distant relatives. These so-called bond groups, in turn, belong to larger groups, called clans.

William Langbauer of the Pittsburgh Zoo and several colleagues, including Charif, have characterized several specific infrasonic calls based on when they occur and how elephants hearing them react. Elephants appear to produce their extremely low-pitched sounds with a larynx similar to those of all mammals, but much larger.

When individual family members reunite after being separated, they greet each other enthusiastically. Excitement increases with the length of time they are separated. They trumpet, scream, and touch each other. They also use a greeting rumble, which begins at a low 18 Hz, crests at 25 Hz—just audible to us—and falls back to 18Hz. An elephant attempting to locate its family uses the contact call, a relatively quiet, low tone with a strong overtone we can hear. Immediately after contact calling, an elephant will lift and spread its ears and rotate its head, as if listening for a response. The contact answer is louder and more abrupt than the greeting, trailing off at the end. Contact calls and responses may continue for hours until the elephant successfully rejoins her family. At the end of a meal, when it's time to move on, one member of a family moves to the edge of the group, typically lifts one leg and flaps her ears. She repeats a "let's go" rumble, which eventually rouses the whole family, who then hit the road.

Unlike the highly social females, males leave their families at about 14 years of age. They travel alone or congregate in small loose groups with other males, occasionally joining a family on a temporary basis. When males come into musth, they wander widely, searching for receptive females.

Females typically come into estrus only once every four years, and then for only four days. So competition is intense and males must have some way of finding mates from long distances. A male in musth repeats a distinctive set of calls called musth rumbles, listening for a response afterward. Males who hear this sound keep away, as bulls in musth are aggressive and dangerous. Females, however, answer with the so-called female chorus, consisting of several females answering with a call similar to the greeting rumble, but somewhat lower. Females will also give this call when a musth male joins their group or when they smell the strong urine of a musth male. A male homes in on the female chorus, hoping to find a female in estrus. After mating, the female rumbles out the post-copulatory sequence, a group of six grunts with strong overtones. She repeats this sequence several times, continuing for up to half an hour.

All of these calls serve as short-range communication in elephants. Documenting the effectiveness of long-range communication has proved technically difficult, however, even among radio-collared elephants. Despite the difficulties, says Charif, "Elephants may routinely know the whereabouts (and maybe activities) of other elephants that are several miles away from them. When a biologist in the field observes the behavior of a group of elephants, s/he may be missing a lot of subtle long-range interactions."

Hippos

Several land mammals other than elephants produce infrasound. But the hippopotamus may be the only one that calls and hears in stereo—one channel in air and a second underwater. Ornithologist William Barklow of Framingham State College in Massachusetts got a shock while observing a river full of hippopotamuses in 1987. Barklow normally studies loons, and his trip to Africa was a holiday. It turned out to introduce him to a new research direction. Sitting near the bank of a river, Barklow saw a hippo emerge from the water. The animal stared silently for a few moments, then suddenly let out a bellow that echoed off the river's banks, shaking Barklow, figuratively and literally. He could actually feel the sound. Hippos can achieve 115 decibels, well into the range of loud, close thunder. Several times during the afternoon Barklow heard a hippo bellow and noticed a curious phenomenon. Distant hippos would surface and bellow back. Had they heard the sound underwater? Could they also vocalize under water?

Intrigued, Barklow studied hippo social communication in libraries and, finding nothing, arranged a study trip to Tanzania to do exploratory work. In Africa, Barklow noticed males defending territories in the river, confirming observations of earlier researchers. He also found that a bellow produced by a male holding a territory often triggered a chorus from other hippos—the ensuing noise traveling along the river and inciting bellows from males at least a mile away. The wave of hippo calls travels up and down river, perhaps conveying information about where the individuals are. The sound might help individuals know when they enter the territory of other hippos and avoid conflicts.

Hippos in the Luangwa River, Zambia.

The shape of a hippopotamus' head aids in what Barklow calls amphibious communication. With its flat upper jaw, upward-pointing nostrils, and top-mounted ears, a hippo can keep its mouth, lower jaw and throat submerged. A bellow bursts forth from the nostrils, accompanied by twin geysers. The airborne sound obviously comes from the nostrils. Yet Barklow also recorded sounds with an underwater microphone, and he could see other hippos surface immediately after he heard both the above- and below-water sounds.

Back in his lab, Barklow analyzed the recordings he had made in the wild. The results both intrigued and frustrated him. Computer analysis showed the sounds he had recorded to be much more complex than he had at first supposed, hinting that hippos possess a more complex communication system than he had assumed. Barklow also found traces of overtones—like those observed in sonograms of elephant calls—indicating the presence of infrasound. Since he had not expected to find

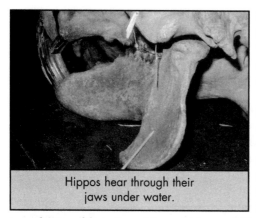
Hippos hear through their jaws under water.

infrasound, Barklow had not brought equipment with him that was capable of recording such low frequencies.

Infrasound makes sense in long-distance communication both in air and in the water, where sound travels more than four times as fast. (The wavelength also increases more than four times, making the sound even better able to travel long distances.) But where did the underwater sound originate? Sound in air bounces off an air-water interface because of the difference in densities, just as sound in water bounces off the water-air interface. The likely path: through a large blob of blubber the hippo carries just under its jaw, acting like the dolphin's fatty forehead melon to conduct and channel sound. The fat, nearly the same density as water, bridges the gap between the hippo's airway and the water.

This neat solution, however, doesn't solve a second problem. How do hippos hear underwater? Two problems arise with ears—flooding and the troublesome water/air interface. To solve the flooding problem, hippos fold back their external ears, sealing off the canals. But this maneuver makes their ears even less sensitive to sound underwater. While sound vibrations in water shake the animal's tissues, and bones can transmit the sound to the inner ear, in general, such a system would be so inefficient and diffuse that it would be unlikely to support a complex underwater communication system. From dolphin researchers Barklow learned that dolphins possess a peculiar jaw that conducts sound clearly and efficiently to the middle ear. Could hippos hear in a similar way?

Studying hippo jaw and ear anatomy convinced Barklow that jaw hearing might work. If so, they might not only hear underwater sounds clearly, but might also hear the underwater sound before hearing airborne sound. This technique could give a floating animal a way to determine the distance of an infrasonic bellow. Sound travels faster in water than air and would reach the middle ear twice, once by way of the jaw and later through the ears—the greater the difference in arrival times, the farther the bellow must have traveled. To gather data on his provocative hippo hypothesis, Barklow returned to Africa once more.

As part of 800 pounds of equipment he lugged halfway across the world on his second research trip in 1992, Barklow brought with him an ingenious way of comparing underwater and airborne sounds, while at the same time monitoring hippo behavior—a stereo video camera connected to an underwater microphone on one stereo channel and a conventional microphone on the other. He also set up an underwater speaker to play back sounds to the hippos.

Barklow returned to Africa during a dry season and he found that low water had increased crowding, which had increased social contacts. Adolescents "argued," sometimes coming to blows, young hippos played and the territorial males kept other males out.

Much of this activity took place underwater and, by recording above and below water, Barklow found that hippos carry on much of their communication underwater. Although the sounds were extremely loud—easily recorded by underwater microphones—they remained

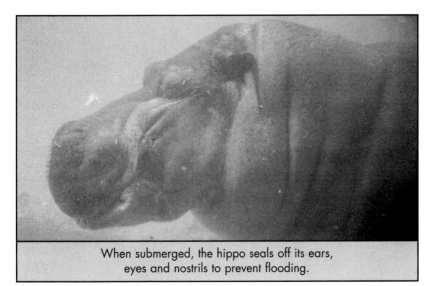

When submerged, the hippo seals off its ears,
eyes and nostrils to prevent flooding.

Infrasonic Calls and How They Work

Pedestrians can hear the thump of bass coming from an over-amplified car stereo while remaining blissfully unaware of the higher pitches. We hear the rumble of distant thunder but can't hear the crack of lightning unless it strikes close by. We hear the bass in these sounds but not the treble because powerful low-frequency sound travels long distances well. But the bass humans hear cuts off at about 20 Hz. Below that, we can only feel the sound at close range, as it vibrates in our chests. Elephants, whales, hippopotamuses, the okapi and rhinoceroses, however, appear to hear and produce sounds well below our range. And it's not just the large animals that hear infrasound. Pigeons, guinea fowl, cod, cuttlefish, octopus, squid and the capercaille, a Eurasian grouse, all hear infrasound.

Sound, as vertebrates hear it, consists of waves of relatively high and low air pressure. When these waves reach the eardrum, they push it in and pull it out, setting in motion a vibration transmitted through the middle ear bones to the cochlea, where specialized cells produce nerve impulses. Our brains interpret these impulses as sound. The waves we hear best range from about a yard peak to peak down to tiny fractions of an inch. Infrasound waves range from tens of yards to miles in length.

These long waves travel through brush and trees relatively

inaudible above the water-air interface. Hippos sometimes call underwater just as they do in air, spewing a double trail of bubbles. But they also apparently squeeze air back and forth between vocal cavities, creating a sort of croak not accompanied by air bubbles. These croaks correlated on videotapes with scenes of calves playing underwater, although Barklow is not sure which animals made the sounds.

In addition to the bellows and croaks, Barklow recorded underwater trains of clicks, similar to those used by killer whales for echoloca-

unimpeded because the ability of an object to reflect a sound wave depends on the ratio of the wavelength to the size of the object. Small objects, such as grass stems, leaves and trees, have no effect on very long infrasound waves. But they reflect and scatter higher frequencies easily. Even the molecules in air absorb a good deal of high-frequency sound, while leaving infrasound unaffected. This makes infrasound ideal for long-distance communication.

Sound travels from its source in all directions, losing about six decibels for every doubling of distance. With no interruption, sound spreads spherically. (Certain shapes can modify the spreading effect. A horn couples the sound source—a trumpet player's lips or the tiny speaker in a horn-loaded audio speaker system—with the air. The larger opening then becomes the sound source, and sound spreads from there.) In a field study of infrasound in Africa, Langbauer found that environments such as bare ground, tall grass and woodland had little or no effect on sounds below 60 Hz projected from a custom-made Pachyderm 2 loudspeaker and recorded at four distances from about 10 to 125 yards from the speaker.

Even very large speakers can't duplicate the power of elephant infrasonic calls, which have been measured at near-thunder levels about five yards from the elephant. The speakers used in field experiments, although huge by home stereo standards, can only produce volumes half that. Extrapolating from playback experiments, Langbauer estimates elephants can hear infrasonic calls at least 2.5 miles away.

tion. Barklow always saw underwater social interactions correlated with clicks, however, so he suggests they serve as communication rather than echolocation. Even stranger sounds, like a Bronx cheer, come from hippos fluttering the valves covering their nostrils. Barklow has no idea how these sounds function. When Barklow played back hippo calls, using his underwater speaker, the hippos oriented themselves toward the recorded sound, bolstering his supposition that hippos use their jaws for hearing and locating underwater sounds.

Cat and Dog Families

The Lion's Roar

With an earsplitting roar that rolls more than three miles across the Serengeti, the "king of the jungle" fluffs his mane, stretches his neck, and asserts domination over his territory. If decades of children's books and Hollywood movies haven't ingrained this name and image, the 1994 Disney animated film hit *The Lion King* did. Never mind that lions don't live in jungles, that a pride doesn't have a king, or that the territory belongs to the females; the image persists.

Prides consist of long-lasting, complex social groups with up to 18 closely related adult females, along with subadults and cubs of both sexes. These females hunt cooperatively, eat cooperatively, and defend and care for cubs cooperatively—even to the extent that any female will nurse any cub in the pride.

Male lions also form cooperative social groups, smaller ones called coalitions. A coalition joins a pride, fathering all the cubs during their tenure. For most males, this temporary residence in a pride represents a once-in-a-lifetime chance to father cubs and pass on their genes. Once a male leaves a pride, he has little chance of finding a place in another. These high stakes cause competition to take on life-and-death importance for male lions, and a coalition remains in a pride only about two to three years before being ousted by a new, more powerful coalition.

One result of this intense competition is the willingness of male lions to attack, and sometimes kill, males intruding on their territory. A more grisly result is that a new regime systematically kills all young cubs and drives away all subadults in a pride. The loss of their young causes females to come into heat sooner, boosting the opportunities for new males to mate. Researchers estimate that attacks by adult male lions cause nearly a third of all cub deaths. (Females rarely attack cubs.) A third result is the formation of coalitions themselves. No one lion, no matter how powerful, can protect his interests alone. He needs partners. Two or three unrelated males may form a coalition, all sharing in fathering cubs. It's a win-win situation wherein males do best by cooperat-

A lion dispute over territory.

ing unconditionally. Larger coalitions form as well, but this decreases the chances of an individual male to father cubs, so larger groups always contain closely related males. If a male doesn't father a cub, at least he protects cubs that are genetically related to him.

Females compete as well, but for long-term ownership of a territory. Encounters with intruding females more typically include a chase, rather than an attack, and the intensity is much lower.

Both male and female lions benefit from knowing exactly who a given lion is. For a male, attacking a member of his own coalition roaring at the boundaries of the territory would weaken his coalition and might end its reign. But attacking one or more intruders from another coalition can make the difference between passing genes to the next generation or not. To make this distinction, lions should be able to distinguish the roar of a friend from the roar of a foe.

Jon Grinnell, Craig Packer and Anne Pusey, all of the University of Minnesota, and Karen McComb, of Cambridge University, decided to find out just how well lions hear and understand roars. Armed with tape recordings of roars of foreign lions, a high-quality playback system, and a stuffed lion, they set out for Serengeti National Park in Tanzania. They studied a population of about 200 lions in 20 prides that have been under observation for 20 years. Each lion bears individual natural marks, making identification possible.

The researchers asked two sets of questions. The first: Can female lions distinguish between the roars of resident males (the fathers of their cubs) and intruders, who might kill their cubs, and can they further distinguish female roars? The second set consist of the two questions: Can males recognize the roars of intruding males, and what are their strategies for dealing with the intrusion?

Many studies of birds and some of fish and mammals have revealed that adults can identify other adults of their own species. But scientists have had a hard time explaining just what the immediate survival benefit of such recognition is. The lion studies provide a potential answer. By recognizing an intruding male lion by his roar, a female could save the lives of her cubs, an immediate genetic benefit. Although she will have new cubs with new males eventually, her best strategy is to ensure that the cubs survive long enough so that invading males will merely chase them away instead of killing them.

African lions mating.

In the study of females, the team set up a loudspeaker in brush more than 200 yards from a pride. They played recorded roars of the fathers of the cubs, roars of male lions living about 20 miles away and unfamiliar to the pride, and roars of unfamiliar female lions. They found females remained calm when they heard the roars of their current mates. (This was the first research, to the team's knowledge, to show that an animal can recognize its mate's call.) But when they heard the roars of unfamiliar males, they would snarl and immediately move toward their cubs to begin a timely retreat. If their cubs were too small to retreat quickly, or were safe in a den, the females stood their ground near the cubs. Pusey and Packer have shown in previous research that the strategy works. Takeovers by new male coalitions rarely succeed unless most of the cubs are too young to travel quickly (under about six months).

Females reacted differently to the roars of unfamiliar female lions. They approached the speaker, sometimes even leaving their cubs behind. But they didn't approach willy nilly. Females apparently assessed the number of females they heard on the tape—the recordings included single roars and choruses of two or three animals—and calculated the odds. They never approached unless the odds were about two to one in favor of the home pride. With lower odds, females either did not move or roared in an apparent attempt to recruit others. When they approached, they did so faster and less hesitantly when the odds were more heavily in their favor.

The researchers also played the recorded roars of a single male lion or from two or three males roaring together. They played roars either from a speaker concealed in brush or in a stuffed male lion. Male lions appear exquisitely aware of exactly who is roaring, clearly distinguishing coalition members from intruders. When faced with a foreigner, males adopted an entirely different strategy from females. No matter what the odds, male lions aggressively approached the roaring speaker, in three cases attacking the stuffed lion. If a new coalition were to succeed in taking over, not only might they lose the cubs they'd already fathered, but also lose any future chances at fatherhood. Males depend on their coalition for their one chance at reproductive success, so no matter what the risk, they cooperate to defend their joint fatherhood (a far cry from the image of a scheming subordinate male plotting to oust the monarch portrayed in *The Lion King*).

The Dog's Bark

First one dog barks, then another four doors away joins in. Soon a whole suburban neighborhood rings with the seemingly endless racket. What are the dogs discussing? Raymond Coppinger and Mark Feinstein, of Hampshire College in Amherst, Massachusetts, think the dogs are being a bit childish.

Most members of the dog family yip, growl and howl. And they share a number of similar facial and body-posture communications. A fully bared set of teeth sends a pretty universal signal, even to us. But the bark truly sets dogs apart. Wolves and coyotes—and some dog breeds, such as the so-called barkless basenji—bark rarely. But some dogs bark incessantly. Coppinger and Feinstein once clocked a dog barking continuously for seven hours. Another scientist recorded 907 barks in a ten-minute period. Coppinger and Feinstein asked why.

Another habit of domestic dogs gives the tip-off as to why domestic dogs bark and perhaps the reason why there are domestic dogs at all.

A dog duet—what does it mean?

That habit is rooting through garbage. Few of us who own dogs have failed to experience finding the kitchen garbage strewn across the floor or the outdoor garbage can upended. Dog fossils occur along with those of humans in archeological sites about 10,000 years old and Coppinger and Feinstein suggest that these dogs were scavenging among humans garbage even that long ago. Humans may well have killed and eaten some of these dogs, but researchers find it highly unlikely that humans domesticated dogs deliberately. Instead, scientists suggest that dogs found an available ecological niche, hanging around human encampments.

The puzzle of the dog's bark comes from comparing it with other animal vocalizations. Low, noisy grating sounds, which we call growls, almost universally suggest adult animals. To us, growls sound dangerous. A dog's or wolf's low menacing growl communicates a threat. A clear, high-pitched whine or yip is unthreatening. A dog whines to come in if it's outside and (sometimes immediately afterward) to go out if it's in. We hear no threat in a whine.

But a bark appears to combine both kinds of sounds, Coppinger and

The wolf's howl communicates over long distances.

Feinstein found, as if the bark conveyed an adult message, harsh and low, along with a youthful keening. The researchers think this fits well with their proposal of domestic dog evolution. Dogs that were least afraid of humans fared better in and near human settlements. But such tameness is a juvenile trait. We can safely handle many baby animals, but the adults of most nondomesticated species are too dangerous to approach. The environment around human encampments selected for tameness. A fearful dog would have a harder time reaching the choicest bits of garbage; a tame one, brave enough to approach the garbage pile, would "fit in."

As the tamer dogs bred, they passed on more of the genes for tameness to subsequent generations and along with tameness came other juvenile characteristics, Coppinger and Feinstein argue: most dogs never develop a fully adult hunting instinct. Mother dogs nurse puppies, but rarely provide solid food for them. Even dogs' variable colors and sizes and the fact that they come into heat twice a year (instead of once as is common in wild canines) relate to the juvenilization, Coppinger and Feinstein propose.

Along with this package of changes came barking, a mixture of juvenile and adult sounds. The barks may make no more sense in the life of a dog than playfulness or floppy ears. They just do it.

Signaling Play

A scene familiar to any dog owner: the dog suddenly folds its front legs, resting on its elbows, head level and eyes aimed up. The rear end, meanwhile, has remained standing, as if it had a life of its own. Like the bark, this communication appears to combine juvenile submission with adult readiness to run. This gesture, common to dogs, wolves and coyotes, communicates play time. Called a bow, the body position may precede a sharp yip toward the human. But does bowing function in canine-to-canine communication? Marc Bekoff, of the University of Colorado, studied videotapes of adult dogs, domestic puppies, wolf pups and coyote pups, noting exactly what behavior followed and preceded each bow. He often saw a bow follow or precede a move that could be interpreted as threatening, such as a bite. Bekoff concludes the bow acts not merely as a stereotyped "Let's play" signal, but as a sort of body language punctuation, attaching the meaning "play" to other actions, and reassuring the playmate that the rules of rough and tumble haven't changed.

Lion cubs play-fight as well, practicing pouncing and other skills they'll need as adults. But even small cubs sport sharp claws and teeth, so they need to communicate that an attack is just for fun. Lion cubs tell their partners they're still playing by walking in a stilted, exaggerated fashion and by keeping their claws sheathed during play.

The dog on the left exhibits a "play bow."

Monkeys

**W**hen an African vervet monkey spies a snake in the grass, it understandably screams. The obvious similarity to a human scream of terror led ethologists for decades to assume the vocalizations carried only the same entirely emotional meaning. But in 1967, Thomas Strusaker, then of the University of California, Berkeley, called that obvious conclusion into question. He described three distinct vervet calls and showed that the call a particular monkey gave depended not on how startled it was, but on what kind of predator it saw. The degree of fear or surprise further modified these calls, but Strusaker could recognize the three calls by sound and by the way other vervets reacted.

A vervet gave one call when it saw a snake. Other members of the troupe then stood on their hind legs and scanned the ground. A second type of call always followed the sighting of a leopard. Other members of the troupe immediately climbed to the smallest branches of nearby trees, safe from the heavy leopard. Lastly, a vervet called yet a third way when it saw a martial eagle cruising the sky. Clinging to the outer branches of a tree, or standing tall in the grass would leave the monkeys vulnerable to this attack. So they stayed near the trunk, deep in the tree, or dove into dense bushes.

Vervet monkeys live in close-knit groups in forests and open areas. Because groups remain stable over long periods, communication between individuals seems valuable to any individual. Strusaker's discovery of three types of calls, which triggered three responses that could be modified by degree, argued for a level of communication beyond a mere shout of fear. But perhaps the responses of troupe members represented merely "monkey see, monkey do." Hearing a snake-call, a monkey might look up, see the calling monkey standing and scanning the grass and just follow suit. Robert Seyfarth, Dorothy Cheney and Peter Marler, all at Rockefeller University at the time, began a series of attempts to manipulate the vervet communication system to answer the question, "Do vervet calls contain information?"

Vervet mother and nursing young.

A vervet stands and scans the area.

Seeing an animal's call as an attempt to refer to an object, rather than as a simple emotional reaction, was a radical departure from earlier research, so the team took great precautions to avoid pitfalls. They recorded many alarm calls and arranged to play them back to a vervet troupe. They spent time with the monkeys, allowing the troupe to become accustomed to their presence. They took care not to play an alarm call when the monkey recorded was in clear sight and obviously not alarmed. They waited until the monkeys were quiet, in no real danger. Lastly, they filmed vervet responses to the recordings.

The monkeys responded to recordings just as they had to original calls. Snake-in-the-grass calls caused troupe members to stand up and scan the ground. Leopard-calls sent them to the farthest reaches of tree limbs and an eagle-call triggered a retreat into bushes or the middle of trees. The first one to respond could not be imitating, since there was no other monkey to imitate. The troupe's response was related to the call itself, not to any real danger, since there was no danger at all. These results, while forcing a dramatic change in ethologists' assumptions about animal calls, did not assign meanings to the calls. A snake-call could mean "snake," or "look on the ground," or may not even have a "meaning" in the sense that human words have meaning.

In a separate experiment, Cheney and Seyfarth played recorded alarm calls of young vervets. Adult females responded by looking at the infant's mother, not at the infant. This behavior implies they can both identify an infant by its distinctive call and can understand the relationship between mother and offspring.

Cheney and Seyfarth went on to analyze the more subtle vervet calls called grunts. Only with practice could humans detect differences in grunts. Earlier research had assumed these differences depended on different situations, not on different meanings. Cheney and Seyfarth tested this assumption, again with playback experiments. Vervets responded differently and consistently to different grunts, no matter what context the researchers chose. When they heard grunts recorded from a dominant male, vervets moved away from the speaker. When they heard grunts of a subordinate to a dominant male, they never moved. Again, the experiments do not assign a meaning to the grunts. But they do show that different grunts convey meaning; they are not all equivalent.

The vervets' social system relies on individuals knowing and communicating social status. And they are not alone among monkeys. Harold Gouzoules and Sarah Gouzoules of Emory University and the Yerkes Primate Center in Atlanta have studied alarm calls made by young rhesus macaque and pigtail macaque monkeys and find that they too communicate information. In a well-studied group of rhesus macaques on Cayo Santiago Island off Puerto Rico, the Gouzouleses tested the long-standing assumption that animal alarm calls vary smoothly, indicating primarily the degree of agitation of the caller.

Because the troupe had been studied for so long, the researchers could identify individuals and knew their family relationships. They knew which animals held dominant rank and which were submissive. They recorded juvenile monkey screams, making voice prints with a sound spectroanalyzer. Contrary to previous assumptions, the vast majority of voice prints fell into one of five recognizable categories, which the researchers called noisy, arched, tonal, pulsed and undulat-

Brain Asymmetries
in Monkeys and Humans

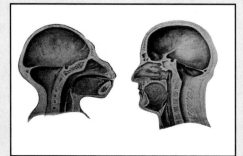

If the monkey's vocal-izations carry a signifi-cant amount of informa-tion referring to objects, in addition to their emo-tional content, can close observation find other language-related similari-ties between monkeys and humans? Marc D. Hauser of Harvard University analyzed videotapes of the Cayo Santiago rhesus monkeys frame by frame, looking for differences in the right and left sides of the monkeys' face. In humans, the right side of the brain controls the left side of the face and vice versa. The right side of the brain appears to concern itself with emotion, while the left side of our brains is concerned with language.

Hauser analyzed four expressions of emotion—a fear grimace,

ing screams. The voice prints varied in frequency and complexity and, although the monkeys appeared to distinguish the sounds easily, the researchers had to rely on voice prints at first.

When the sound prints were compared with videotapes of monkey behavior, researchers found that monkeys used each scream category in a specific social situation. Noisy screams indicated that an aggressor of higher rank had made physical contact. Relatives of a screaming juve-nile reacted appropriately to a higher ranking monkey, attempting to distract the aggressor rather than confront him directly. Undulating screams also told of an attack by a higher ranking opponent, but with-out physical contact. Arched screams indicated a lower ranking aggres-

which a subordinate monkey gives when being attacked or intimidated by a higher ranking group member; a copulation grimace, similar to the fear grimace, but more fleeting; and two expressions given by a dominant animal toward a lower ranking one: an open mouth threat with lips in an O shape and an ear-flap threat, with the ears flat against the head.

Hauser found that in most monkeys, the left side of the face tends to move earlier, takes on a more extreme expression and retains the expression longer than the right side.

These results may reveal that in rhesus monkeys, as in humans, the right hemisphere of the brain dominates emotional expression.

Adding Hauser's evidence to previous studies of Japanese macaques, which showed the left side of the brain to be involved in perception of vocal signals, leads to the suggestion that both humans and monkeys have the same pattern of brain asymmetries related to referential vocalization and facial expression of emotions.

Hauser cautions, however, that this conclusion rests on the assumption that monkey vocalizations carry more symbolic information than emotional. Secondly, he points out that rats and chickens also show some brain asymmetries associated with communication, so these asymmetries may not indicate an increased ability to communicate symbolic information.

sor and did not indicate any physical contact. Pulsed and tonal screams tended to indicate a squabble within the immediate family. The screams appeared to function as recruitment calls, communicating specific information about the situation and the location and identity of the calling individual, not just the degree of fear. In addition to telling listeners who was screaming, the juveniles communicated who was attacking, giving relatives information needed to respond appropriately. As with vervets, rhesus macaques react to recorded calls the same way they do when they hear live calls. Harold Gouzoules likens the screams to simple words, functioning as symbolic representations.

Can We Talk?

Humans have probably always recognized a family resemblance in the great apes. The name we use for the great tree-dwelling, red-haired apes of Borneo comes from the Indonesian *orang*, person, and *hutan*, jungle. A few people have kept young chimps as pets. Many have owned plush toy chimps to cuddle at night. The famous chimpanzee researcher Jane Goodall traces her fascination with these great apes to a toy chimp named Jubilee, which she kept for decades.

At the same time, we have sought to distance ourselves from the beasts, often using language as the defining difference. In the first century B.C., the Roman historian Sallust wrote, "All men who would surpass the other animals should do their best not to pass through life silently like the beasts..." In the 1600s, Descartes found a universal human truth in "I think, therefore I am." But animals, Descartes declared, didn't think; they were mere automata, beast machines. Descartes' follower, La Mettrie, however, pointed out that deaf people have a difficult time learning to speak and speculated that with the right teacher, a chimpanzee could learn and thereby become "a little gentleman."

This schizophrenic attitude persists today. No reputable scientist disputes Darwin's assertion of physical continuity from the simplest animals to humans. Great apes clearly share much with humans. Their anatomy resembles ours more than any other mammal, and even their brains have similar—though smaller—parts. If researchers could emulate the fictional Dr. Doolittle and converse with an animal, surely that animal would be a great ape. But some scientists disagree. Despite a continuity of other traits, they say, language stands alone, not merely the most complicated kind of communication, but a unique capability, unattainable by any non-human animal.

Early attempts, from the 1900s through the 1930s, to teach chimps to speak met with dismal failure, vindicating the critics. The animals just couldn't wrap their otherwise expressive lips around words. In the most successful cases, they made sounds charitably interpreted as short words,

An orangutan's hoot.

Nim Chimpsky and trainer Joyce Butler practice American Sign Language.

such as "mama," "papa," "cup" and "up," after years of training.

Following La Mettrie's suggestion that a gifted teacher of the deaf could succeed with chimpanzees, a 1925 scientific article suggested sign language as an alternative. But serious efforts to teach non-vocal communication to apes only began in the 1960s. Researchers attempted to teach individual signs derived from American Sign Language (ASL) to Washoe, a chimpanzee; Koko, a gorilla; and Chantek, an orangutan. Sarah, a chimpanzee, learned to manipulate arbitrary plastic symbols standing for words, and another chimpanzee, named Lana, used an early computer keyboard, with arbitrary symbols the researchers called lexigrams.

All these projects succeeded where earlier speech projects had failed. The apes learned to use hand gestures, plastic symbols or keyboards to communicate with their trainers. The 1960s and 1970s became the golden age of ape language-learning. Researchers claimed (and some continue to claim) that the apes had learned tens or even hundreds of signs. But popular accounts went farther—that apes held conversations and had "learned sign language." To this day, assertions that apes can converse with humans using symbols or sign language abound in popular magazines and books and even college textbooks.

But although the trained apes often used more than one sign or symbol in a sequence and could clearly get a message across—most often a request for food or attention—researchers wondered if the apes had learned Language, with a capital L. Some researchers working in the field feel justified in using the word "language" to describe the results of these experiments, but Steven Pinker, psycholinguist and author of *The Language Instinct*, disagrees. Pinker has declared, "No chimpanzee has learned sign language...They've certainly learned some gestures, but sign language is not just a system of gestures. It's a full, grammatical language with its own systematic grammar, like Latin." Setting the idea of a full language aside, however, did the apes' hand gestures constitute words? Did they truly understand that signs or lexigrams stood for objects or actions? Were their strings of two or three signs sentences?

Herbert Terrace of Columbia University attempted to solve the sentence problem with a chimpanzee called Nim Chimpsky, named in humorous honor of Noam Chomsky, the renowned linguist. Chomsky had asserted that if apes could use language, they would do so in the wild. They don't, he said, so they can't. Terrace taught Nim gestures based on ASL and succeeded, just as others had succeeded in teaching Washoe, Sarah, Koko, Chantek and Lana. With near round-the-clock help from a platoon of students, Terrace signed with Nim, recording long strings of signs and accumulating a huge body of data on audio tape and film.

Terrace's teaching schedule and what he called the baby-sitting problem brought Nim's tenure at Columbia to a close. Nim moved on to the Institute for Primate Studies in Norman, Oklahoma and Terrace turned to an extensive analysis of the data gathered during Project Nim. In his 1979 book *Nim*, Terrace wrote: "The regularities in our corpus that were noted before Nim returned to Oklahoma gave me reason to believe that Nim was creating primitive sentences. Our intensive post-Oklahoma effort at data analysis had hardly begun, however, when I began to doubt that Nim's combinations were legitimate sentences."

Terrace concluded that despite long strings of signs such as "give orange me give eat orange me eat orange give me eat orange give me you," Nim's actual "sentences" averaged 1.5 signs. Had Nim's learning ground to a halt, or had he just gotten what he wanted without longer strings of signs? Terrace concluded that Nim had never signed a true sentence and that many of Nim's individual signs immediately followed similar signs by his trainers. Close examination of his films convinced Terrace that Nim mostly imitated trainers, often after prompting. Furthermore, Terrace analyzed films of other ape projects, including two about Washoe, and concluded that in those projects, too, trainers repeatedly prompted and then interpreted separate responses as sentences.

In one example, Washoe is supposed to have signed "Baby in my drink," when shown a doll and a cup. Terrace describes the film: "Washoe is with her teacher Susan Nichols, who has a cup and a doll. Ms. Nichols points to a cup and signs 'that'. Washoe signs 'baby'. Ms. Nichols brings the cup and the doll closer to Washoe, allowing her to touch them, then slowly pulls them away, signing 'that' and pointing to the cup. Washoe signs 'in' and looks away. Ms. Nichols brings the cup and doll closer to Washoe again, who looks at the two objects once more and signs 'baby.' Then, as Ms. Nichols brings the cup still closer, Washoe signs 'in'. 'That,' signs Ms Nichols, and points to the cup. 'My drink,' signs Washoe. Now the question is, is this utterance by Washoe—'baby in baby in my drink'—either spontaneous or a significant, creative use of words?"

Terrace reluctantly concludes, "Until it is possible to defeat all plausible explanations short of the intellectual capacity to arrange words according to a grammatical rule, it would be premature to conclude that a chimpanzee's combinations show the same structure evident in the sentences of a child." Terrace softened this stern proclamation with, "This is not to say that a chimpanzee is simply not capable of creating a sentence."

This statement notwithstanding, other researchers in ape language-learning saw Terrace's conclusions as an attack. Because of Terrace's well-reported change of mind, funding dwindled for projects, some researchers left the field, others distanced themselves from the scientific community, and a sense of blame persists. In the 1994 book,

Animal Learning and Cognition, Duane M. Rumbaugh and E. Sue Savage-Rumbaugh write, "But the view of Terrace and his associates prevailed to the extent that it soon became widely accepted that (1) because Terrace's Nim did not have language and (2) because analyses at Terrace's laboratory of taped materials from the laboratories of other signing projects indicated that others' apes were also imitating, there was the strong implication that (3) no ape had demonstrated any language competence whatsoever and that (4) language was beyond the competence of apes!"

Rumbaugh and Savage-Rumbaugh run one of the few animal-language research projects to continue full steam after Terrace's change of mind, at Georgia State University's Language Research Center. Following up on the Lana project, Rumbaugh, Savage-Rumbaugh, Rose A. Sevcik and others continued teaching language to great apes, with some remarkable results and some disappointments. Two chimps, Sherman and Austin, learned lexigrams for foods and for simple tools

Nim spontaneously signs "hug" to his teddy bear, held by H.S. Terrace.

they could use to get the food from closed containers. With extensive training, they then learned to cooperate, using keyboards to ask each other for the tools necessary to obtain food, which they shared.

Later they learned lexigrams for the category food and the category tool. With little training, they accurately categorized 20 lexigrams for food and 20 lexigrams for tools—a significant linguistic feat in itself. But another task hinted at things to come. With no specific training, both chimps were able to look at a lexigram, then reach into a box they couldn't see into and pull out the named object. These feats appeared to show at least that Sherman and Austin grasped the concept of naming. They seem to use symbols as words.

But a separate project with an adult ape called Matata turned out to be a bad-news-good-news situation. Matata belongs to a species called the bonobo (*Pan paniscus*), a close relative to the chimpanzee (*Pan troglodytes*). Often called—incorrectly—pygmy chimpanzees, bonobos differ from chimpanzees in several respects. They stand a head shorter, weigh less and have a more gracile body shape. They stand upright more often, have different faces and even communicate in the wild differently. Furthermore, their social structures in the wild differ from those of chimpanzees. Some see these differences as making bonobos more like humans. Could they learn language more easily? The bad news was Matata failed to reliably use even a few lexigrams after years of training.

Matata was a working mom, however, and brought her adopted baby Kanzi to work. And therein lies the good news part of the tale. While Matata sat bemused by the keyboard, Kanzi crawled in her lap and on her back or played nearby. The researchers tolerated Kanzi, but never trained him. He grew up for two years in an environment where humans continuously made sounds to his mother and tapped at a keyboard, trying to teach her individual signs. Savage-Rumbaugh finally gave up on Matata—not only hadn't she learned to use lexigrams to ask for what she wanted, but she had a dangerous tendency to roughly grab from researchers. They allowed Kanzi to remain at the Language Research Center and sent Matata back to the nearby Yerkes Primate Center where she could use her native communication mode to find a mate. Kanzi was two and a half years old.

Like the child of an immigrant, Kanzi soon showed he had absorbed just what Matata had resisted. Within a week, he began to use the keyboard to make his desires known spontaneously. But he also appeared to name objects, even when he did not want the object. Savage-Rumbaugh and Sevcik decided not to train Kanzi at all, but to see if he could continue to soak up the keyboard "language" during daily interactions with researchers, who talked to him, using both lexigrams and speech, as if he understood. In other words, they treated Kanzi the way parents treat a preverbal child constantly hearing language. Kanzi's keyboard helped in this effort by generating synthesized speech to sound out the English word for each lexigram.

Made wary by the Terrace incident, Savage-Rumbaugh and her colleagues attempt to meet the many legitimate objections raised by Terrace and others. They try to avoid prompting, and to construct experiments that will stand up to close scrutiny. Furthermore, the researchers are careful with their claims. They use the phrase "non-random lexigram combinations" instead of "sentences," for example. But Kanzi works for attention, not food, and the team can't eliminate people from the experiments. And Kanzi's non-random lexigram combinations rarely exceed three lexigrams. Watching Kanzi in casual "conversation" with Sue Savage-Rumbaugh, an observer is struck by the intense give and take, reminiscent of Terrace's description of the "baby in my drink" film clip.

Kanzi's two-and-three-word sentences on the keyboard may seem less than impressive. But a set of experiments comparing Kanzi's understanding of spoken English to that of Alia, the two-and-one-half-year-old daughter of a Language Center researcher, appears to show a very different level of understanding. Kanzi and Alia were presented with sentence-understanding tasks as similar as the researchers could make them.

Archival videotape of Kanzi's performance sets the scene. Kanzi sits in a room with two researchers (one is Rose Sevcik). A third (Sue Savage-Rumbaugh) stands outside the room with a microphone. The two researchers inside wear earphones playing loud music to reduce the chance they can give Kanzi clues. The room has a "kitchen" and a large playroom with a number of objects Kanzi has never seen. A child's toilet, a pitcher of water, a rubber snake, a stuffed dog, a 25-pound bag of carrots, a hand puppet vaguely resembling a rabbit.

The voice from outside says "Kanzi, make the dog bite the snake." Kanzi immediately picks up the rubber snake and the plush toy dog. He carefully puts the snake's head into the dog's mouth and gently squeezes the dog's jaws shut. An impressive show of understanding made more impressive by the fact that Kanzi has generalized the spoken words dog and snake to toys he's never seen.

"Kanzi, tickle Rose with the bunny," says Savage-Rumbaugh. Kanzi picks up a bunny hand puppet, carries it to Sevcik and tickles her. Sevcik says in explaining the videotape that Kanzi's only previous knowledge of "bunny" was a videotape of a Language Research Center worker dressed in a bunny suit. The researchers had never drilled

Mother orangutan and her infant communicate nonverbally.

Kanzi (or Alia) on the requests and all the objects were new, purchased just for the experiment.

Duane Rumbaugh summarizes the results: "Kanzi's comprehension of 500 novel sentences of request were very comparable to Alia's. Both complied with the requests without assistance on about 70 percent of the sentences." He emphasizes that Kanzi learned by observation alone very early in life and that the researchers discovered this fact only by the lucky decision to keep Kanzi around after Matata was sent home. "The apes can come to understand even the syntax of human speech at a level that compares favorably with that of a two-to-three-year-old child—if they are reared from shortly after birth in a language-structured environment....Reared in this manner, the infant ape's brain develops in a manner that enables it to acquire language. First through its comprehension and then through its expression, a pattern that characterizes the course of language acquisition in the normal child....We had no intention of studying language-observational learning in (Kanzi). But it happened and we've replicated it with other (bonobos and chimpanzees)," Rumbaugh says.

Duane Rumbaugh and Sue Savage Rumbaugh summarize—fairly, it seems—the current state of research in their chapter of the 1994 *Animal Learning and Cognition*. "Though none will argue that any animal has the full capacity of humans for language, none should deny that at least some animals have quite impressive competencies for language skills, including speech comprehension."

Lost in the Brouhaha: Apes' Own Communication

Almost lost in arguments about whether and to what degree apes can learn "language" are their own native "tongues." Each species uses a number of varied—and sometimes incredibly loud—vocal calls, as well as facial expressions and a rich body language. No researcher claims that these calls constitute a language as complex as ours. But they serve the animals perfectly well in the wild.

Orangutans may have the most impressive individual call of any ape, the long call. It begins with a low soft grumble, modulating in pitch like a string bass player using vibrato. It builds slowly to a roar audible a mile and a quarter away through the dense Borneo jungle. The third part falls back to a soft series of mumbles and sighs. As with many animal calls, this one announces a male, protecting his territory and possibly calling females. Some orangutans accompany this call by pulling over standing snags, creating a crash that echoes through the forest. Occasionally, the crash alone is enough to trigger return long calls from neighboring males. More often, males will call back and forth to each other, presumably communicating their location and the sovereignty of their territory.

Orangutans also vocalize with grunts, grumbles and squeaks when they copulate, and young orangs squeak, bark and scream. Adults and young make a variety of sounds with their lips and throats, sucking, burping and even grinding their teeth. Researchers have no dictionary of these sounds, or the gestures and body postures orangutans use to communicate. Because they live high in the trees of dense forest, wild orangutans have proved difficult to document.

What chimpanzees lack in individual volume compared to orangutans, they make up for in cacophonous chorus. Jane Goodall and many other researchers have catalogued a wide variety of screams, grunts and so-called pant-hoots, all accompanied by striking facial expressions and body language.

Chimpanzees live in fluid social groups that change in both the short and long term. Unlike dolphins, elephants and lions, male chimpanzees form stronger and longer-lasting social bonds than do females. They cooperate in grooming and hunting, forming alliances that increase their social rank and mating success.

One of the best-studied of chimpanzee vocalizations, the pant-hoot, begins with breathy, low-pitched hoots that segue into a series of quicker, higher-pitched in-and-out pants, as if the chimp were trying to play harmonica without an instrument. Finally, the pant-hoot builds to a loud, climactic crescendo. Both sexes pant-hoot and appear to do so at every opportunity when it's appropriate to express excitement. Males and females pant-hoot differently and even humans can discriminate between the pant-hoots of individuals with a little practice. Chimpanzees listen to distant pant-hoots and respond. These sounds may serve as identification, but they occur in such a wide variety of circumstances and with enough subtle variations that they may well carry other meanings as well.

Male chimpanzees sometimes accompany long-distance pant-hoots by drumming with their hands or feet on tree buttresses or hollow stumps or logs, and appear to use pant-hoots to communicate their location over long distances. Recent field studies tend to indicate that males keep in touch with specific individuals, primarily allies.

At least two groups of chimpanzees may speak different dialects. Well-studied groups at Gombe and Mahale pant-hoot slightly differently. Because long-distance identification of allies and rivals carries such importance for chimpanzees, dialectical differences in isolated groups makes sense. But dialects in birds, for example, arise as fledglings learn songs from their nearest neighbors. No one has yet shown that chimpanzees learn their pant-hoots from adults. Differences in habitat or in the genetic makeup of the groups could also account for the different pant-hoots.

Chimpanzees in Zaire.

Threads from a Tangled Skein

ut of a tangled skein of observations of animal communication, some common strands emerge. The profound effect of human language on human thought positively leaps out. Humans—no matter where we live—acquire an incredibly complex language so early in life that it colors everything we think. In fact, memories of events that occurred before we had words to describe them remain hazy and vague—if we can remember any such early events at all. And we are so steeped in our ability to symbolize reality that we tend to forget the difference between the symbol and the reality, confusing, for example, "my country" for the place we live and "the enemy" for the fellow human at the other end of the battlefield.

It's no wonder, then, that we think of animal communication in terms of human language. We tend to measure animals' ways of communicating against the "standard" of human language. Scientists, linguists and philosophers argue about whether apes can truly communicate in human sign language or with arbitrary symbols organized in a humanlike grammar. But we know relatively little about apes' natural modes of communication.

A tendency to see human verbal language as all important shows up in the story of Clever Hans, the "mathematical" horse. People of the time stood ready to believe that Hans could understand human speech and carry out verbal commands to add and subtract—accomplishments observers could identify with. But they declared Hans a fraud and lost interest when skeptics revealed that the horse was "only" exquisitely sensitive to human body language—a means of communication of which we are only vaguely aware.

We are similarly impressed with a dog that seems to understand it's master's words, but uninterested in his truly amazing ability to identify the three neighborhood dogs who have recently visited his favorite lamp post from the odor of their urine.

Humans harbor an acute conscious awareness of communicating. But do animals? A wide spectrum of conscious awareness exists between the extremes of an unconscious knee-jerk reflex and composing a sonnet, but scientists and philosophers have laid animal commu-

Kanzi holds in his mouth a stone flake that he made.

nication at each end. Less extremely, Steven Pinker argues persuasively in *The Language Instinct* that animal communication represents far less than the equivalent of human language, and in *Animal Minds*, Donald R. Griffin argues persuasively that animal communication represents far more than a reflex. A safe stance, then, would call animal communication neither a poem nor a twitch. But placing animal communication more precisely on the spectrum may prove difficult. Even in the case of talking parrots, the animals best able to imitate human language, and the great apes, our nearest relatives and probably the "smartest" non-human animals, we know little about how conscious they are of communicating, and researchers do not agree on how much their behavior resembles human language.

Another thread to emerge from a study of animal communications tells of the power of evolution. Given an ecological niche, the slow (on biological time scale), random process of evolution will fill it. If males expend great energy to attract females, they open an ecological niche for "cheaters," males who sit quietly and wait for the female to arrive. The presence of such sneaky males itself opens a niche for females with some means of identifying the true owner of the territory. If niches become available for a wide variety of bird species in a tropical forest, not only will new species arise, but they will evolve unique means of communication.

The tremendous variety of modes of communication that have evolved demonstrates the fundamental role of communication among all animals. Animals use every sense, gesturing with appendages and body position; sending and receiving subtle—or not so subtle in the case of frightened skunks—odor signals; squeaking, squawking, singing and chirping; sending and receiving electrical signals; flashing lights; changing skin pigmentation; "dancing;" and even tapping and vibrating the surface they walk on.

The full range of data collected by animal communication researchers, though vastly broader than what this book touches on, still only represents the tip of the iceberg. No doubt future research will bring a richer understanding of those instances of animal communication we know about today, and will bring to light many new and equally fascinating stories of communication among the animals.

Watching the passing scene.

Abbreviations: PR = Photo Researchers

p. 2—© Susan Kuklin/PR

p. 3—© Lawrence Naylor/PR

p. 4—© Mary Thacher/PR

p. 11—© Susan Kuklin/PR

CHAPTER 1
*p. 23—© Gregory Ochocki/PR p. 24—© American Museum of Natural History
p. 25—© George Lower/National Audubon Society/PR p. 27—© Gregory Ochocki/PR*

CHAPTER 2
p. 29—© Scott Camazine/PR p. 31—© Stephen Dalton/PR

CHAPTER 3
p. 34—© Jerome Wexler/PR p. 36—© J. H. Robinson/PR p. 38—© Stephen Dalton/PR

CHAPTER 4
*p. 41—© J. H. Robinson/PR p. 42—© G. Uetz p. 43—courtesy G. Uetz
p. 45—© P.J. Watson*

CHAPTER 5
*p. 47—© Ron Church/PR p. 48—© Tom McHugh/PR p. 49—© Tom McHugh/ National
Audubon Society/PR p. 50—© Tom McHugh/PR p. 52—courtesy Rasnow and Assad/Caltech
p. 53—courtesy Andrew Bass, photo by Margaret Ann Marchaterre*

CHAPTER 6
*p. 55—courtesy Mac F. Given p. 56—courtesy Mac F. Given p. 57—© Leonard Lee Rue/
National Audubon Society/PR p. 58—© A. W. Ambler/National Audubon Society/PR
p. 59—© Jen and Des Bartlett/PR*

CHAPTER 7
*p. 61—American Museum of Natural History p. 63—American Museum of
Natural History p. 66—© Graham Pizzey/PR*

CHAPTER 8
*p. 71—© Francois Gohier/PR p. 72—© Rick Sternback/PR p. 77—© Francois Gohier/PR
p. 79—© James Aroyan from the Journal of the Acoustical Society, Nov. '92
p. 80—© Francois Gohier/PR*

CHAPTER 9
*p. 83—© Peter S. Thacher/PR p. 84—© George Daniell/PR p. 87—© Ian Cleghorn/PR
p. 88—courtesy Barklow p. 89—courtesy Barklow*

Worker honeybees on brood comb.

I N D E X

Gerald Borgia "Sexual Selection in Bowerbirds," *Scientific American*, June 1986.

Charles Darwin *The Origin of Species and the Descent of Man*, Modern Library, New York, 1977.

P. J. DeVries "Singing Caterpillars, Ants and Symbiosis," *Scientific American* 267, 1992.

Frans B.M. de Waal *Chimpanzee Politics (Power & Sex Among Apes)*, Harper & Row, New York, 1982.

Donald R. Griffin *Animal Minds*, The University of Chicago Press, 1992.

David W. Pfennig and Paul W. Sherman "Kin Recognition," *Scientific American*, June 1995.

Steven Pinker *The Language Instinct: How the Mind Creates Language*, HarperCollins, New York, 1994.

Sue Savage-Rumbaugh *Kanzi: The Ape at the Brink of the Human Mind*, John Wiley & Sons, New York, 1994.

Herbert S. Terrace and Eyre Methuen *Nim: A Chimpanzee Who Learned Sign Language*, Alfred A. Knopf, New York, 1979.

Jane van Lawick-Goodall *In the Shadow of Man*, Dell, New York, 1971.

A C K N O W L E D G M E N T S

Many people contributed to the writing of The Language of Animals.
To all of you, my heartfelt thanks:

My wife, Karen, teacher of biology at Peninsula College, and my eleven-year-old son Robbie, a budding naturalist, writer or rock bassist, (depending on when you ask him), for uncounted dinner-table conversations about animal communication, and for many helpful suggestions.

My friend and fellow science writer Carolyn Strange, for many years of moral support and excellent writing advice.

All the scientists who took time to bring to light and explain the research that forms the basis of The Language of Animals, *not only for making the book possible, but for enriching my own view of the animal world.*

Rose Sevcik, for showing me the Language Research Center of Georgia State University, and introducing me to Kanzi the bonobo, owned by Yerkes Primate Research Center, Emory University.

James Ha, of the University of Washington, for posting my query on the animal behavior electronic mail network he administers. As a result, I received communications from dozens of researchers, many of whose work appears in this book. (My animal communication electronic mailbox contains more than 600 messages.) Without the wonder of electronic mail, research for this book would have taken months longer.

The several university and institutional public information officers who suggested and helped me contact researchers, especially Cathy Yarbrough of the Yerkes Primate Research Center, Emory University.

The Language of Animals *represents a synthesis of thousands of pieces of information in the form of research papers, telephone interviews with researchers, first hand observations, and hundreds of electronic mail messages, as well as popular books, text books, popular magazine articles and videotaped TV specials. The result is a communication from me to the reader. I alone must take responsibility for mistakes of fact or interpretation.*

— STEPHEN HART